FROM DIRAC TO NEUTRINO OSCILLATIONS

FROM DIRAC TO NEUTRINO OSCILLATIONS

by

Tino Ahrens
Georgia Institute of Technology,
Atlanta, Georgia U.S.A.

KLUWER ACADEMIC PUBLISHERS
Boston / Dordrecht / London

Distributors for North, Central and South America:
Kluwer Academic Publishers
101 Philip Drive
Assinippi Park
Norwell, Massachusetts 02061 USA
Telephone (781) 871-6600
Fax (781) 871-6528
E-Mail <kluwer@wkap.com>

Distributors for all other countries:
Kluwer Academic Publishers Group
Distribution Centre
Post Office Box 322
3300 AH Dordrecht, THE NETHERLANDS
Telephone 31 78 6392 392
Fax 31 78 6546 474
E-Mail <orderdept@wkap.nl>

Electronic Services <http://www.wkap.nl>

Library of Congress Cataloging-in-Publication Data

Ahrens, Tino.
From Dirac to neutrino oscillations / by Tino Ahrens.
p. cm.
Includes bibliographical references and index.
ISBN 0-7923-7886-5 (acid-free paper)
1. Quantum theory. 2. Transformations (Mathematics) 3. Particles (Nuclear Physics) I. Title.

QC174.17.T7 A37 2000
530.12--dc21

00-041595

Printed on acid-free paper.

Printed in the United States of America

FOR MARTIN

CONTENTS

PART TWO

Preface

This text is meant to be a view of the quantum mechanical formalism as it develops with the successive introduction of different types of transformations. In particular, it is meant to help the readers with three tasks: acquainting themselves with a general and direct approach to the quantum mechanics of spin-one-half and spin-one particles, primarily leptons, photons and massive vector bosons, and to some extent quarks; finding out what some of the related areas of current research interest are; and, last and foremost, trying to understand the subject, beginning with and stressing the principles involved. The exposition is based on finite-dimensional representations of the homogeneous Lorentz group, and the subsequent introduction of gauge transformations, of the Abelian and non-Abelian varieties. Reference to classical mechanics is avoided. Acting on the simple basis spinors and vectors, Lorentz transformations generate wave and field functions. Equations are obtained by the relativistic generalization of the addition of angular momenta, the wave or field functions being the solutions. For zero mass the equations may be obtained as the limits of the equations for the massive cases or by the application of the Euclidian group in two dimensions. The latter approach is illuminating in that it uncovers a loss in generality resulting from the former. Identifying momenta as eigenvalues of translations demonstrates covariance under the inhomogeneous Lorentz or Poincare group.

Various representations of wave and field functions are given. Particularly in the case of the Dirac equation, one representation may be better suited to a given problem than others. Spin-one fields are treated as 6-vectors and as antisymmetric tensors. 6-vectors are particularly obliging, when it comes to the description of polarization. Occasionally reference is made to quantum field theory, as in the case of the interpretation of free particle wave functions of negative energy, or when one extracts wave functions from the integral expression of a potential. This does not require a preceding course in quantum field theory.

The interaction with a "massless potential" is established by the demand of local gauge invariance. A discussion is added of the behavior of an interacting "massive potential" as the mass approaches zero. The behavior of the electromagnetic radiation potential under Lorentz transformations is developed, and compared to that of the electromagnetic field in the narrow sense of the word. Remarks are made concerning the formal distinction between fields and potentials.

The foundation of the material up to here is well understood, which is not supposed to mean that a deeper foundation does not exist. The same cannot be said for the rest of the material. The foundation is spread out by extending Abelian to non-Abelian local gauge invariance. It is then deepened by including the self-interacting scalar field pioneered by Goldstone and elaborated on by Higgs, imposing on it too the extended gauge invariance, and then letting it generate mass. But the unique character of this field among interacting fields is not understood. The outstanding result of these ideas, the electroweak theory of Weinberg, Salam and Glashow, is discussed, and the relations of coupling constants and masses, including the Fermi constant, are derived.

The account highlights important milestones, occasionally referring to early works concerning concepts like isospin, charge independence, neutral currents, lepton number, baryon number and others, concepts which have long since become part of our daily bread. The emphasis continues to be on transformations, which at times are sufficiently complex to be given only in infinitesimal form.

The neutrino problem is the ideal context in which to introduce particle-state transformations. They are invoked in the interpretation of neutrino oscillations and double-beta decay. These phenomena are of practical use in the exploration of very small masses, and should eventually permit a decision between Dirac and Majorana neutrinos. The formalism is also of wider interest. Reference is made to the Cabibbo angle, the Kobayashi-Maskawa matrix and the early work of Gell-Mann and Pais. Glashow's theory on the superposition of massive vector-bosons falls into the same general category. Here such diverse and fundamental items as mass matrices, coherence and phase factors of improper transformations are discussed.

Throughout, an effort has been made to refer to the original literature, whether it pertains to theories, improved formulations, experiments or ideas for experiments. Reference to early books is made, whenever the treatments of interest seem excellent, newer ones only being mentioned, if improvements appear significant, or different points of view are offered. Results are given of convincing first experiments, e.g., the famous experiments of Goldhaber, Grodzins and Sunyar and of Davis, Harmer and Hoffman. Later data quoted are mostly from the late nineties, a time when neutrino oscillations became generally accepted. While data will inevitably change, the types of experiments are likely to remain the same for quite some time. So are essentially most of the input data, such as fluxes and cross sections. A few heroic first experiments are mentioned, foremost among them the Cowan-Reines experiment, even if they did not - or at least at first did not - yield decisive data.

This text is mainly intended for the more theoretically inclined physics students with at least three years of serious college physics under their belts. Because of the points of view offered and some of the detailed descriptions, it may also be of interest to more advanced workers in the field. It might serve as a conceptually slightly demanding introduction to relativistic quantum mechanics and to mathematically simpler aspects of particle physics, or as a follow up to a standard course in relativistic quantum mechanics, hopefully leading to a mathematically somewhat deeper understanding of the subject. - So as not to interrupt the continuity of the development, background material is presented in the appendix.

Strong interactions are not treated in this book - well, almost not. In the discussion of isospin strong interactions figure prominently. A condensed recapitulation of the historical development is the natural way to see the changing relevance of what is now called “strong isospin”. It is about energy regimes and unfolding structure. Some texts tend to leave the uninitiated in the dark about this. The late blossoming of the concept of “weak isospin” can be attributed to the difficulty of experiments with weakly interacting neutral lepton-currents. The rather long lasting, wide spread conviction of the nonexistence of such currents did not help. As elsewhere in the text, this attempt at not shortchanging history too flagrantly is also meant to have its practical side. Points of departure may offer themselves, and so may analogies useful for further developments.

A course of study with the aid of this text should lead to a more thorough understanding of some of the elements in texts like Bjorken and Drell’s “Relativistic Quantum Mechanics”, or Sakurai’s “Advanced Quantum Mechanics”. Thus Bjorken and Drell obtain the Dirac equation in the manner of Dirac himself, demanding, to achieve a positive definite probability density, an equation linear in the time derivative, and so, because of Lorentz covariance, also linear in the gradient. Subsequently, Lorentz covariance is explicitly imposed on the equation. Sakurai starts by showing how van der Waerden’s relativistic second-order equation may be obtained by generalizing the nonrelativistic equation for the two-component spinor of a spin-one-half particle. He proceeds to demonstrate the equivalence of this second-order equation with two first-order equations, coupled by the particle’s mass, the Dirac equation. These approaches are clearly very physical, however quite specific. While more mathematical, the approach taken here is more general. Lorentz covariance is the starting point. The Dirac equation is but one massive-particle equation resulting from the connection by space and time derivatives of two subspaces of different representations of the homogeneous Lorentz group. Because of the particular representations involved, it is a very special equation. Maxwell’s equations are obtained by the same scheme. - Because of the two subspaces connected by them, the massive counterparts to Maxwell’s equations are second-order equations. To obtain a first-order equation for massive spin-one particles, one needs to connect three subspaces. The result is the Kemmer equation.

The lectures presented here in book form were never intended as a standard introduction to relativistic quantum mechanics. The concern here is not with teaching how to calculate, say, relativistic scattering problems. There have been excellent books around for quite some time accomplishing that very purpose, just two of which I mentioned above. Rather, our concern is with wave functions, (unquantized) field functions, the equations they satisfy for various “elementary” particles, the interactions of the latter, including relations between their strengths, and, most importantly, how all of these may be thought to come about.

In thinking about a general subject for decades, one obviously draws on many sources, some of which, although continuing to be represented in one's thoughts, are no longer identified. With this in mind, my formulations of Part One have perhaps been influenced most by Corson's "Introduction to Tensors, Spinors and Relativistic Wave Equations" and Weinberg's lectures on "The Quantum Theory of Massless Particles" at the 1964 Brandeis Summer Institute. As already pointed out in the beginning, the approach here is from the bottom up. Lagrangians and Hamiltonians are only introduced towards the end. The famous Noether Theorem is given in the Appendix. There the relation between Lorentz invariance and charge conservation is pointed out. It is built on free fields!

Acknowledgements

My thanks to Frank Avignone for keeping me informed of recent developments in experimental neutrino physics, in particular of his and his collaborators' work in double-beta decay; to Johan Belinfante for explaining work of his on Lie groups; to Helmut Biritz for a long and fruitful association, with so many conversations about particle physics; to N. N. Chakravorty for discussions related to the zero-mass limit; to Ed Lerner for enlightening discussions of fundamental differences between spin-one-half and spin-one fields; to Bruce McFarland for freely bouncing a plethora of ideas back and forth; to Fred Reines for conversations about his groups' work and ours on neutral currents, theoretical and experimental, and for kindly lending me his support in work on massive photons; to Harald Stumpf for stimulating conversations about quantum field theory, and for steadily communicating to me the results of his iconoclastic work; to James Whittenton for keeping me supplied with an abundance of current neutrino-oscillation data and their interpretation, including his own ideas; and finally to John Wood for valuable pointers concerning isospin and CP nonconservation. Special thanks to Audrey Ralston for patiently typing and retyping the manuscript, and to Martin Ahrens for drawing the figures.

PART ONE

Spinors and vectors, insofar as they are of interest in quantum mechanics and field theory, not including gravity, are formulated as they result from the application of Lorentz transformations. Equations of physical interest which they satisfy are given as consequences rather than starting points. Covariance properties of these equations, and of their solutions and associated quantities, are formulated. Expressions for massive and massless particles and fields are carefully distinguished. A few relevant experiments are described. - A basic knowledge of special relativity, nonrelativistic quantum mechanics, in particular the theory of angular momentum, and the fundamentals of electromagnetism will be helpful when following this approach. Acquaintance with a modicum of group theory, while not necessary, may hasten the understanding of some passages. Some group theoretical help is provided in the Appendix.

Chapter 1

COORDINATE AND POINT TRANSFORMATIONS

Consider a scalar function of space and time as it undergoes a coordinate transformation,

$$\psi(x^{\mu}) = \psi(x^{\mu}(x'^{\nu})) = \psi'(x'^{\mu}) \,;\; \mu,\nu = 1,2,3,0 \,;\, x^{0} = ct \;. \tag{1a}$$

An observer who uses primed coordinates sees as $\psi'(x'^{\mu})$ a situation described in unprimed coordinates by $\psi(x^{\mu})$. One can also write

$$\psi'(x^{\mu}) = \psi(\tilde{x}^{\mu}) \tag{1b}$$

i.e., instead of a transformation t on x , giving x', one uses t on $\tilde{x}$, giving x. The transformation

$$\psi(x^{\mu}) \rightarrow \psi'(x^{\mu}) \tag{2}$$

is a "point transformation", as distinct from a coordinate transformation. Eq. (1a) tells us that a point is fixed; the coordinates describing it change. Utilizing eq. (1b), the transformation (2) changes the point; the coordinates are fixed.

Instead of a function of coordinates, we may use a function of 4-momentum components, $\psi(p^{\mu})$, p^{0} = E/c. The functions may themselves be components, which transform along with the variables. Since in the following we only need linear transformations, we may write

$$\psi'^{m}(x^{\mu}) = t^{m}_{\;\;n}\, \psi^{n}(\tilde{x}^{\mu}) \tag{3a}$$

where

$$\tilde{x}^{\mu} = t^{-1\mu}_{\quad\;\nu}\, x^{\nu} \tag{3b}$$

and similarly for $\psi^m(p^\mu)$ and p^μ. Script indices generally cover a different set of numbers than Greek indices. Also, the Einstein summation convention is used. - For the time being we restrict ourselves to homogeneous transformations.

Chapter 2

ROTATIONS AND LORENTZ TRANSFORMATIONS

Describe a coordinate (not point) rotation about the third axis of a Cartesian coordinate system by

$$\begin{aligned} x'^1 &= x^1\cos\varphi + x^2\sin\varphi \\ x'^2 &= -x^1\sin\varphi + x^2\cos\varphi \ . \end{aligned} \tag{4a}$$

If for positive angles the rotation is clockwise when looking along the axis of rotation towards plus infinity, the coordinate system is right handed. Describe a Lorentz transformation by

$$\begin{aligned} x'^1 &= (x^1 - vx^0)/\surd(1-v^2) \\ x'^0 &= (-vx^1 + x^0)/\surd(1-v^2) \ . \end{aligned} \tag{4b}$$

For positive speeds, here along the first axis, we call it a "standard Lorentz transformation". For these examples $\psi'(x^\mu)$ may be produced from $\psi(x^\mu)$ (note eq. (1b)) by infinite repetition of the infinitesimal transformations

$$1+\varepsilon(x^1\partial/\partial x^2 - x^2\partial/\partial x^1) \ and \ 1+\varepsilon(x^1\partial/\partial x^0 - x_0\partial/\partial x^1) \ . \tag{5}$$

We recognize the first quantity in parenthesis as the orbital-angular-momentum operator L_3, multiplied by i. The second quantity in parentheses, divided by i, we call M_k. All in all there are six such independent "generators" $L_{\mu\nu}$, where

$$L_{kl} \equiv L_m \ and \ L_{k0} \equiv M_k ; \ k,l,m = cycl.1,2,3 \ .$$

We adhere to the metric where

$$x^k = x_k , x^0 = -x_0 \ .$$

(For the general meaning of sub and superscripts see A.10). Furthermore, we use natural units,

$$\hbar = c = 1 \ .$$

As has been well known since the beginning of quantum mechanics, the L_k satisfy the commutation relations

$$[L_k, L_l] = iL_m \ . \tag{6a}$$

The remaining relations are

$$[M_k, M_l] = -iL_m \ and \ [L_k, M_l] = iM_m \ . \tag{6b}$$

The finite dimensional, irreducible matrix representations of the abstract operators satisfying these commutation relations will be denoted by J_k and K_k [A.2]. The "unitarian trick" (Weyl 1946)

$$J_k = A_k + B_k, K_k = i(B_k - A_k) \tag{7}$$

leads to quantities, A_k and B_k, which separately satisfy the angular-momentum commutation relations, the A_k commuting with the B_k. A finite-dimensional irreducible representation of the "proper homogeneous Lorentz group", L_h, the group of all rotations and Lorentz transformations, is then usefully denoted by (A,B), with A and B the maximum values of A_3 and B_3, respectively. The third component of the angular momentum is customarily taken to be the diagonal one, and, on account of eqs. (7), it is indicated to adhere to this tradition, when generalizing from three-dimensional space to space-time.

Chapter 3

SPINORS

The lowest dimensional non-trivial representations are (1/2,0) and (0,1/2). For the non-zero A_k and B_k we take one half times the Pauli sigmas (Schiff 1949),

$$\sigma_1 = \begin{pmatrix} 0 & 1 \\ 1 & 0 \end{pmatrix} \quad \sigma_2 = \begin{pmatrix} 0 & -i \\ i & 0 \end{pmatrix} \quad \sigma_3 = \begin{pmatrix} 1 & 0 \\ 0 & -1 \end{pmatrix} . \tag{8}$$

We form infinitesimal transformations, like

$$1 + i\varepsilon\sigma_3/2 \quad and \quad 1 - i\varepsilon(\pm i\sigma_1)/2 \tag{9a}$$

which for positive ε correspond to clockwise rotations about the third axis and standard Lorentz transformations along the first axis, respectively. In the second of eqs. (9a) the upper sign holds for transformations in (1/2,0), the lower for those in (0,1/2). From the recipe [A.12]

$$\lim_{n\to\infty} (1 + x/n)^n = \exp(x)$$

the finite transformations

$$\exp(i\sigma_3\varphi/2) \quad and \quad \exp(\pm\sigma_1\alpha/2) \tag{9b}$$

follow. Let the matrices of eqs. (9) multiply spinor components, in analogy to eqs. (4), which can rather obviously be regarded as the result of the multiplication by matrices of the Cartesian and time components of an event. The spinor components, like the components of an event, are then "contravariant" [A.10].

Consider the "basis spinor"

$$\xi_u = \begin{pmatrix} 1 \\ 0 \end{pmatrix}$$

u standing for spin-up along the third axis. To obtain a general "standardized spinor" we follow convention (Weinberg 1965) and apply to ξ_u the matrix

$$T_B = R^{-1}(\theta,\varphi) B_B(\alpha) R(\theta,\varphi) \tag{10}$$

where the "boost" $B_B(\alpha)$, which, as we shall see shortly, increases the momentum, is taken along the third axis. The subscript B implies the representation (0,1/2). This subscript should therefore also be attached to the operated on spinor. With the rotation in "spin-space"

$$R(\theta,\varphi) = \begin{pmatrix} \cos\theta/2 & \sin\theta/2 \\ -\sin\theta/2 & \cos\theta/2 \end{pmatrix} \begin{pmatrix} \exp(i\varphi/2) & 0 \\ 0 & \exp(-i\varphi/2) \end{pmatrix} \tag{11a}$$

and the boost

$$B_B(\alpha) = \begin{pmatrix} \exp(\alpha/2) & 0 \\ 0 & \exp(-\alpha/2) \end{pmatrix} \tag{11b}$$

the resulting spinor, which we choose to call $\xi_{B1/2}$, is

$$T_B \begin{pmatrix} 1 \\ 0 \end{pmatrix}_B = \begin{pmatrix} \exp(\alpha/2)\cos^2\theta/2 + \exp(-\alpha/2)\sin^2\theta/2) \\ (\exp(\alpha/2) - \exp(-\alpha/2))\exp(i\varphi)\sin\theta/2\cos\theta/2 \end{pmatrix} . \tag{11c}$$

Standardizing means here producing a well defined correlation between momentum and spin, see eqs. (12). Also, we know there are two two-by-two representations of L_h corresponding to the familiar four-by-four representation of special relativity, namely (1/2,0) and (0,1/2). So far we have just used the latter. ξ_{Bu} is a member of the subspace belonging to it.

3.1 MASSIVE PARTICLES

One physical interpretation is to let the up-spinor be the spin function of a particle at rest. For the exponential

$$\exp(\alpha) = \surd(1+v)/\surd(1-v)$$

we then get

$$\exp(\alpha) = (p^0+p)/m$$

where $m, v, p, and\, p^0$ are the mass of the particle, and its speed, momentum and energy after the boost. Expressing the functions of polar angle, azimuth and boost parameter in terms of Cartesian momentum components and energy

$$p^1 = psin\theta \cos\varphi\ ,\ p^2 = psin\theta \sin\varphi\ ,\ p^3 = pcos\theta$$

gives for $\xi_{B\,1/2}$

$$\sinh(\alpha/2) \begin{pmatrix} \dfrac{\surd((p^0+p)/m)(p+p_3)/2p + \surd(m/(p^0+p))(p-p_3)/2p)}{(p^0+p-m)/2\surd(m(p^0+p))} \\ \\ (p_1+ip_2)/p \end{pmatrix} \qquad (12a)$$

$$= k \begin{pmatrix} p^0+m+p_3 \\ p_1+ip_2 \end{pmatrix} = \xi_{B\,1/2}(p^\mu)$$

where k equals $\sinh(\alpha/2)/p$. Similarly one obtains with the spin-down basis spinor

$$\xi_{Bd} = \begin{pmatrix} 0 \\ 1 \end{pmatrix}_B$$

$$T_B \xi_{Bd} = k \begin{pmatrix} p_1 - ip_2 \\ p^0 + m - p_3 \end{pmatrix} = \xi_{B-1/2}(p^\mu) \ . \tag{12b}$$

Using the transformation T_A, which differs from T_B only in the boost, and thus basis spinors belonging to (1/2,0), we obtain

$$T_A \xi_{Ad} = k \begin{pmatrix} -(p_1 - ip_2) \\ p^0 + m + p_3 \end{pmatrix} = \xi_{A-1/2}(p^\mu) \tag{12c}$$

$$T_A \xi_{Au} = k \begin{pmatrix} p^0 + m - p_3 \\ -(p_1 + ip_2) \end{pmatrix} = \xi_{A\,1/2}(p^\mu) \ . \tag{12d}$$

The subscripts 1/2 and -1/2 are the J_3 eigenvalues, s, of the original basis spinors.

Be it emphasized, that the momentum components in the second pair of the just obtained spinors are produced by the same boost as those in the first pair. The difference in signs is due to the different boosts the spinor components experience. In two dimensions, the boost generators K_k have two different representations; see eqs. (7) and (9).

It may have been noticed that

$$\xi_{A-1/2} = -i\sigma_2 \xi^*_{B\,1/2} \ and \ \xi_{A\,1/2} = i\sigma_2 \xi^*_{B-1/2} \ .$$

(The difference in signs may be avoided by defining ξ_{Au} with a minus sign). This was to be expected from eqs. (9). Spinor transformations brought about by matrices like those in eq. (9b), let us call them S_A and S_B, respectively, clearly satisfy

$$S_A\, \xi_{A-1/2} = -i\sigma_2\, S_B^*\, i\,\sigma_2\,(-i\sigma_2)\, \xi^*_{B\,1/2}\ .$$

More along these lines in Section 3.1.3.

3.1.1 Wigner Rotations

One may wish to subject the standard spinors of eqs. (12) to an arbitrary Lorentz transformation, S, where by Lorentz transformation we now mean, and shall continue to mean unless specifying otherwise, any transformation in the Lorentz group:

$$\xi^{m}_{C\,s}(p^{\mu}) \rightarrow \xi'^{m}_{C\,s}(p'^{\mu}) = S^{m}_{C\,n}\,\xi^{n}_{C\,s}(p^{\mu})\,,\ C = A\ or\ B\,,\ p'^{\mu} = l^{\mu}_{\nu}\,p^{\nu} \qquad (13a)$$

where the matrix (l^{μ}_{ν}) corresponds to $(S^{m}_{C\,n})$. Furthermore, we may wish to express, if possible, the new spinor in terms of the familiar standard spinors, see eqs. (12), with the new 4-momentum as argument.

For this purpose it is opportune to introduce the rotation (Wigner 1939)

$$R_W = T^{-1}(p'^{\mu})\,S\,T(p^{\mu})\ . \qquad (14a)$$

In our application $T(p^{\mu})$ is either one of the transformations in eqs. (12). S is then S_A and S_B, respectively. Next, we combine the basis spinors ξ_u and ξ_d in a matrix whose elements we call $\xi^{m}{}_{s}$, s denoting the spinor, and *m* the component. In our simple case *m* and s assume the values ±1/2. - In general, when dealing with the representation (A,B), the matrix corresponding to our matrix $(\xi^{m}{}_{s})$ is not square. All that matters is that A_k, B_k and J_k, J_3 corresponding to our s, satisfy the first of eqs. (7). - The above rotation shall be sandwiched between the matrix $(\xi^{m}{}_{s})$ and its inverse, resulting in the "Wigner rotation"

$$\xi^{-1\,r}{}_{m}\,\left(T^{-1}(p'^{\mu})\,S\,T(p^{\mu})\right)^{m}{}_{n}\,\xi^{n}{}_{s} = t^{r}{}_{s}\ . \qquad (14b)$$

The emphasis is on the switch from the script indices *m* and *n*, specifying components, to the indices r and s, specifying basis functions, which are here the basis spinors ξ_u and ξ_d. Thus we have here

$$\xi^n{}_s = \delta^n{}_s \; .$$

The Wigner rotation then is

$$(T^{-1}(p'^\mu)ST(p^\mu))^r{}_s = t^r{}_s \; . \tag{14c}$$

Multiplying eq. (14b) by $\xi^p{}_r$, summing over r, and furthermore multiplying by $T^m{}_p(p'^\mu)$ and summing over the index *p*, yields

$$S^m{}_n \xi^n{}_s(p^\mu) = \xi^m{}_r(p'^\mu) t^r{}_s \tag{14d}$$

always keeping in mind that ultimately we need to include the subscripts A or B. This equation holds quite generally for massive particles, i.e., existing rest frame.

Before proceeding to a nontrivial application of the last two equations, we look at the rotation about the third axis: we apply eq. (14d) to the representation (0,1/2),

$$S^m_{B\,n} \xi^n_{Bs}(p^\mu) = \xi^m_{B\,r}(p'^\mu) t^r{}_s \tag{15a}$$

and put

$$S_B = \begin{pmatrix} \exp(i\varphi''/2) & 0 \\ 0 & \exp(-i\varphi'') \end{pmatrix} \tag{15b}$$

with $\xi_{Bs}(p^\mu)$ from eqs. (12a) and (12b). When we expressed the rotation angles θ and φ of the spinor arguments in terms of momentum components, we used the standard relations between Cartesian and spherical coordinates. This explains the sequence of factors $R^{-1}(\theta,\varphi)B(\alpha)R(\theta,\varphi)$ used before; see eqs. (11) and (12). Remember, for positive angles, $R(\theta,\varphi)$ produces clockwise rotations; see eqs. (4a). For s = 1/2 eqs. (15a) and (15b) result in

$$\begin{pmatrix} \exp(i\varphi''/2) & 0 \\ 0 & \exp(-i\varphi''/2) \end{pmatrix} \begin{pmatrix} p^0+m+p_3 \\ p_1+ip_2 \end{pmatrix} = \begin{pmatrix} p^0+m+p_3 & p'_1-ip'_2 \\ p'_1+ip'_2 & p^0+m-p_3 \end{pmatrix} \begin{pmatrix} t^{1/2}_{\ 1/2} \\ t^{1/2}_{\ 1/2} \end{pmatrix} . \qquad (15c)$$

The matrix t can be obtained from eqs. (10) and (14c). The only change in going from $T_B(p^\mu)$ to $T_B(p'^\mu)$ is in the rotation, where we go from $R(\theta,\varphi)$ to $R(\theta,\varphi')$. Consequently,

$$t = R^{-1}(\theta,\varphi')B^{-1}(\alpha)\, R(\theta,\varphi')\, S(\varphi'')\, R^{-1}(\theta,\varphi)\, B(\alpha)\, R(\theta,\varphi) \ .$$

We know that in three dimensions

$$p^k = R^{-1k}{}_3(\theta,\varphi)p$$

and likewise for the primed quantities.

$$\varphi' = \varphi - \varphi''$$

follows. Thus

$$t = R(0,\varphi'') \ . \qquad (15d)$$

The right side of eq. (15c) becomes

$$\begin{pmatrix} p^0+m+p_3 & p'_1-ip'_2 \\ p'_1+ip'_2 & p^0+m-p_3 \end{pmatrix} \begin{pmatrix} \exp(i\varphi''/2) \\ 0 \end{pmatrix} = \exp(i\varphi''/2) \begin{pmatrix} p^0+m+p_3 \\ p'_1+ip'_2 \end{pmatrix} . \qquad (15e)$$

On to an actual Lorentz transformation for S in eq. (15a):

$$\begin{pmatrix} \exp(\alpha''/2) & 0 \\ 0 & \exp(-\alpha''/2) \end{pmatrix} \frac{\cosh(\alpha'/2)}{\cosh(\alpha/2)} \begin{pmatrix} \cosh\alpha + 1 + \sinh\alpha\cos\theta \\ \sinh\alpha\sin\theta \end{pmatrix} =$$

$$\begin{pmatrix} \cosh\alpha' + 1 + \sinh\alpha'\cos\theta' & \sinh\alpha'\sin\theta' \\ \sinh\alpha'\sin\theta' & \cosh\alpha' + 1 - \sinh\alpha'\cos\theta' \end{pmatrix} \begin{pmatrix} t^1{}_1 \\ t^2{}_1 \end{pmatrix} . \tag{16a}$$

In this equation we expressed $\xi_{B1/2}(p^\mu)$, $\xi_{B1/2}(p'^\mu)$ and $\xi_{B-1/2}(p'^\mu)$ in terms of boosts and rotation angles. To avoid needless complication, we equated the azimuth to zero and the mass to one. Furthermore, the Lorentz transformation does not take the primed momentum out of the 3-1 plane.

A similar equation holds with $\xi_{B-1/2}(p^\mu)$ on the left side, see eq. (12b), when on the right the spinor $t^m{}_2$ replaces $t^m{}_1$. (Instead of denoting the spinor indices by ±1/2 we have used 1 and 2.) Call that equation (16b). Starting from p^0 equal to m, the primed momentum is either reached in two steps, or in one, see Figure 1,

$$\begin{pmatrix} \sinh\alpha\sin\theta \\ 0 \\ \sinh\alpha\cos\theta\cosh\alpha'' + \cosh\alpha\sinh\alpha'' \\ \sinh\alpha\cos\theta\sinh\alpha'' + \cosh\alpha\, cosh\alpha'' \end{pmatrix} = \begin{pmatrix} \sinh\alpha'\sin\theta' \\ 0 \\ \sinh\alpha'\cosh\theta' \\ \cosh\alpha' \end{pmatrix} . \tag{16c}$$

Let us verify the relations

$$t^2_1 = -t^1_2, and\, t^1_1 = t^2_2 = \sqrt{(1-(t^1_2)^2)} \tag{17a}$$

which must hold if t is a (one parameter) rotation matrix. From eq. (16a) follows, under assumption of the first two eqs. (17a),

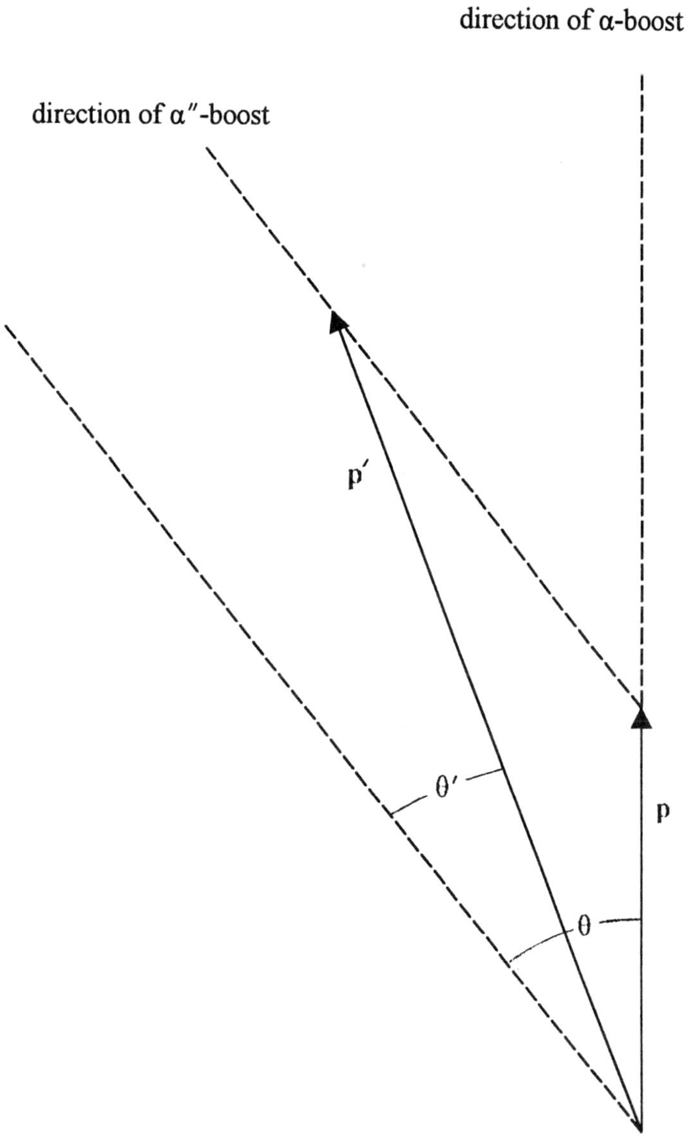

Figure 1: the momentum $\mathbf{p}'$ *is reached in two steps with boosts* α *and* α''*, or in one with boost* α'*; see equations (16a) and (16c).*

$$t^{1}_{2} \cosh\alpha'/2 = \sinh\alpha''/2 \sinh\alpha/2 \sin\theta$$
$$t^{1}_{1} \cosh\alpha'/2 = (\cosh\alpha''/2 \cosh\alpha/2 + \sinh\alpha''/2 \sinh\alpha/2 \cos\theta) . \qquad (17b)$$

Squaring and adding results indeed in the third of eqs. (17a). With cosh(α′/2) from eq. (16c), eqs. (17b) permit evaluation of t.

3.1.2 Connection of Spinors and 4-Vectors

To find out what first order equations the spinors of eqs. (12) satisfy, we first need to look in some detail at the momentum 4-vector, p^{μ}. Since we know from special relativity that a 4-vector is made up of a 3-vector and a scalar in 3-space, and from quantum mechanics that a vector transforms like a quantity of spin-one, and a scalar like one of spin-zero, the first of eqs. (7) suggests that a 4-vector belongs to the representation (1/2,1/2). More succinctly, it suggests that the second-rank spinors belonging to the direct product of two spin-one-half representations, one acting in "A-space" and one in "B-space", transform like the components of a 4-vector. The direct product of two matrices shall be defined by

$$M \otimes N = \begin{pmatrix} m_{11}n_{11} & m_{11}n_{12} & \cdots & m_{12}n_{11} & m_{12}n_{12} & \cdots \\ m_{11}n_{21} & m_{11}n_{22} & \cdots & m_{12}n_{21} & m_{12}n_{22} & \cdots \\ \vdots & \vdots & & \vdots & \vdots & \\ m_{21}n_{11} & m_{21}n_{12} & \cdots & m_{22}n_{11} & m_{22}n_{12} & \cdots \\ m_{21}n_{21} & m_{21}n_{22} & \cdots & m_{22}n_{21} & m_{22}n_{22} & \cdots \\ \vdots & \vdots & & \vdots & \vdots & \end{pmatrix} . \qquad (18)$$

In the case of matrices of the representation (1/2,1/2) M is, e.g., $\exp(i\sigma_2\theta/2)$ or (the boost) $\exp(-\sigma_3\alpha/2)$, while N, using the same parameters, is consequently $\exp(i\sigma_2\theta/2)$ or $\exp(\sigma_3\alpha/2)$, respectively. The "generating matrices" of our direct-product-transformations,

$$(\sigma_k/2) \otimes 1 \; \textit{and} \; 1 \otimes (\sigma_k/2)$$

are for k = 1, 2, 3

$$\frac{1}{2}\begin{pmatrix} 0 & 0 & 1 & 0 \\ 0 & 0 & 0 & 1 \\ 1 & 0 & 0 & 0 \\ 0 & 1 & 0 & 0 \end{pmatrix}, \frac{1}{2}\begin{pmatrix} 0 & 0 & -i & 0 \\ 0 & 0 & 0 & -i \\ i & 0 & 0 & 0 \\ 0 & i & 0 & 0 \end{pmatrix}, \frac{1}{2}\begin{pmatrix} 1 & 0 & 0 & 0 \\ 0 & 1 & 0 & 0 \\ 0 & 0 & -1 & 0 \\ 0 & 0 & 0 & -1 \end{pmatrix} \tag{19a}$$

and

$$\frac{1}{2}\begin{pmatrix} 0 & 1 & 0 & 0 \\ 1 & 0 & 0 & 0 \\ 0 & 0 & 0 & 1 \\ 0 & 0 & 1 & 0 \end{pmatrix}, \frac{1}{2}\begin{pmatrix} 0 & -i & 0 & 0 \\ i & 0 & 0 & 0 \\ 0 & 0 & 0 & -i \\ 0 & 0 & i & 0 \end{pmatrix}, \frac{1}{2}\begin{pmatrix} 1 & 0 & 0 & 0 \\ 0 & -1 & 0 & 0 \\ 0 & 0 & 1 & 0 \\ 0 & 0 & 0 & -1 \end{pmatrix} \tag{19b}$$

respectively. The order from top to bottom of the second rank spinnors $p^{m,n}$, the components which the matrices (19) multiply, follows from eq. (18) and is $p^{1/2,1/2}$, $p^{1/2,-1/2}$, $p^{-1/2,1/2}$, $p^{-1/2,-1/2}$. The eigenvalues of $J_3 = A_3 + B_3$ are obviously 1, 0, 0, -1. This does not mean we already have the decomposition into 3-vector and scalar, but we shall once we form the symmetric and antisymmetric combinations of $p^{1/2,-1/2}$ and $p^{-1/2,1/2}$. The matrix which accomplishes this and also places the three vector components next to one another is

$$D = \begin{pmatrix} 1 & 0 & 0 & 0 \\ 0 & 1/\sqrt{2} & 1/\sqrt{2} & 0 \\ 0 & 0 & 0 & 1 \\ 0 & 1/\sqrt{2} & -1/\sqrt{2} & 0 \end{pmatrix}. \tag{20a}$$

Performing the similarity transformations

$$D(\sigma_k \otimes 1/2 + 1 \otimes \sigma_k/2)D^{-1} \; and \; -D(i\sigma_k \otimes 1/2 - i\,1 \otimes \sigma_k/2)D^{-1}$$

results in the standard generating matrices appropriate to the components $-(p_1-ip_2)/\sqrt{2}$, p_3, $(p_1+ip_2)/\sqrt{2}$ and p_0 (see also Section 4, eq. (70)). Therefore, to relate the $p^{m,n}$ base to the p_μ base is the same as relating $-(p_1-ip_2)/\sqrt{2}$, $(p_3+p_0)/\sqrt{2}$, $(p_3-p_0)/\sqrt{2}$ and $(p_1+ip_2)/\sqrt{2}$ to p_μ. The second-rank spinor components are therefore obtained from the Cartesian components by application of the matrix

$$\Sigma = 1/\sqrt{2}\begin{pmatrix} -1 & i & 0 & 0 \\ 0 & 0 & 1 & 1 \\ 0 & 0 & 1 & -1 \\ 1 & i & 0 & 0 \end{pmatrix} . \tag{20b}$$

We call the columns $\bar{\sigma}^{\mu}$ and rearrange the elements in each of them to form a matrix,

$$\bar{\sigma}^1 = \begin{pmatrix} -1 & 0 \\ 0 & 1 \end{pmatrix}, \bar{\sigma}^2 = \begin{pmatrix} i & 0 \\ 0 & i \end{pmatrix}, \bar{\sigma}^3 = \begin{pmatrix} 0 & 1 \\ 1 & 0 \end{pmatrix}, \bar{\sigma}^0 = \begin{pmatrix} 0 & 1 \\ -1 & 0 \end{pmatrix} . \tag{21a}$$

We now have the invariant association

$$p^{m,n} = \sigma^{\mu m,n} p_{\mu}/\sqrt{2} \tag{21b}$$

$$l^{\mu}{}_{\nu} S^{m}{}_{Ap} S^{n}{}_{Bq} \sigma^{\nu p,q} = \sigma^{\mu m,n} \tag{21c}$$

$\mu, \nu = 0$ to 3; $m, n, p, q, = 1/2, -1/2$. The bar is only used on an entire matrix, when the spin superscripts do not appear, and thus cannot specify the matrix′ form. We shall soon see how going from superscripts to subscripts changes the sigmas.

3.1.3 The Dirac Equation

With the 4-momentum in the form $p^{m,n}$ and the spinors correspondingly written $\xi^{m,0}$ and $\xi^{0,n}$, we can directly combine them in such a way that two one-half-spins add up to zero-spin, first in A-space and then in B-space:

$$\begin{aligned} p^{1/2,1/2}\xi^{-1/2,0} - p^{-1/2,1/2}\xi^{1/2,0} &= -\kappa\xi^{0,1/2} \\ p^{1/2,-1/2}\xi^{-1/2,0} - p^{-1/2,-1/2}\xi^{1/2,0} &= -\kappa\xi^{0,-1/2} \\ p^{1/2,1/2}\xi^{0,-1/2} - p^{1/2,-1/2}\xi^{0,1/2} &= \kappa'\xi^{1/2,0} \\ p^{-1/2,1/2}\xi^{0,-1/2} - p^{-1/2,-1/2}\xi^{0,1/2} &= \kappa'\xi^{-1/2,0} \; . \end{aligned} \tag{22}$$

The signs of the constants on the right have been chosen for convenience. In conformity with the meaning established following eqs. (19), the first superscripts are the eigenvalues of $\sigma_3/2$ in A-space, the second those in B-space. Substituting the explicit $p^{m,n}$ from before, these equations read

$$-\sigma_k p_k \xi_A + \kappa\xi_B = p^0 \xi_A \;\; and \;\; \sigma_k p_k \xi_B + \kappa'\xi_A = p^0 \xi_B \tag{23}$$

where the σ_k assume the form of the Pauli sigmas, and $\xi^{m,0} \equiv \xi_A^m$, $\xi^{0,n} \equiv \xi_B^n$. Evaluation of eqs. (23) results in $\kappa = \kappa' = m$.

In analogy to coordinates which are contravariant, we have all along been using spinors with superscripts. - The eqs. (22) are manifestly covariant under Lorentz transformations, covariant meant here in the general sense and not as distinguished from contravariant. Eqs. (23) are written in matrix form, and consequently the suppressed script summation index of the σ_k, and of the implied unit matrices, must be a subscript. Subscripts specify covariance, as distinguished from the contravariance of superscripts [A.10].

- This paragraph and the next provide an analysis of the connection between the quantities appearing in eqs. (22) and (23). Analyses like this one will not be given for similar equations derived later on. - The transition from superscript to subscript is direct when going from the second pair of eqs. (22) to eq. (23a), for it is the second superscript on the $p^{m,n}$ which is lowered, or $\sigma^{\mu m,n}$ goes into $\sigma^{\mu m,}{}_n$, allowing invariant matrix multiplication; see eqs. (21). The index lowering is accomplished by the "metric spinor", τ, defined by a bilinear form which remains invariant under our spinor transformations. Taking spinors in B-space, we have

$$\xi'^m_B \tau_{mn} \chi'^n_B = \xi^m_B \tau_{mn} \chi^n_B \equiv \xi^m_B \chi_{Bm} \equiv \xi_{Bm} \chi^m_B \; . \tag{24a}$$

The spinor transformations are "unimodular", i.e., they have determinant plus 1. Consequently,

$$\xi^m_B \tau_{mn} \chi^m_B = -\xi^{1/2}_B \chi^{-1/2}_B + \xi^{-1/2}_B \chi^{1/2}_B \tag{24b}$$

and

$$\tau = \begin{pmatrix} 0 & -1 \\ 1 & 0 \end{pmatrix} . \tag{24c}$$

Lowering the second script index on σ^u

$$\sigma^{\mu m,}{}_n = \tau_{np} \sigma^{\mu m,p} \tag{25}$$

results in negative Pauli sigmas and minus σ^0, permitting to explain also formally the appearance in eq. (23a) of σ_k and the unit matrix multiplying p^0: it is due to the substitution of the σ^μ in the second pair of eqs. (22) by minus $\sigma^\mu \tau$. The metric spinor then changes the relative sign of the spinor components, delivering eq. (23a), (multiplied by minus one); σ^μ is the matrix $(-\sigma^{\mu m}_n)$,.

In the case of the first pair of eqs. (22), the first script index on $p^{m,n}$, see eqs. (21), need be lowered, and the matrix transposed. From eqs. (9) we see that the transformations in (1/2,0) and (0,1/2) are related by

$$S_B = S_A^{-1\dagger} . \tag{26a}$$

Thus, spinor components ξ^m_A *and* ξ^*_{Bm} transform the same way. Adopting the convention that a dot over an index implies complex conjugation (Corson 1953), we express this behavior as

$$\xi^m_A \sim \xi_{B\dot{m}} . \tag{26b}$$

Since the various sigmas are unique constant matrices for completely defined representations, we can equate

$$\sigma^{\mu m,}{}_n = \pm \sigma^\mu_{B\dot{m}n} . \tag{26c}$$

Since both indices on the right refer to B-space, they are not separated by a comma. For definitness, we shall use the upper sign in Eqs. (26c). If we then choose

$$\sigma^{\mu}_{m,n} = +\sigma^{\mu\dot{m}}_{B\ n} \tag{26d}$$

the metric spinor for the spinor space belonging to (1/2,0), τ_A, will be the inverse of the one for the space belonging to (0,1/2); see eqs. (24). We may now also equate

$$\sigma^{\mu m,n} = \sigma^{\mu m}_{A\ \dot{n}} \ and\ \sigma^{\mu m,}_{\quad n} = \sigma^{\mu m\dot{n}}_{A} \tag{26e}$$

and consequently

$$\sigma^{\mu}_{B\dot{m}n} = \sigma^{\mu m\dot{n}}_{A} \ and\ \sigma^{\mu\dot{m}n}_{B} = \sigma^{\mu}_{A\dot{m}n} \ . \tag{26f}$$

We evaluate

$$\begin{aligned}(\sigma^{k}_{A\dot{m}n}) &= \sigma^{k}\ Pauli = -(\sigma^{k}_{B\dot{m}n})\\ (\sigma^{0}_{A\dot{m}n}) &= -1 = (\sigma^{0}_{B\dot{m}n})\ .\end{aligned} \tag{26g}$$

Thus, substituting the $\sigma^{\mu m,n}$ in the first pair of eqs. (22) by $\tau^{mp}_{A}\,\sigma^{\mu n}_{p,}$, transposing and letting the metric spinor change the relative sign of the ξ_A components, results in eq. (23b). So much for the formal aspects of eqs. (23).

From here on out, the spinor transformation property of any sigma will be described by the position and dotted or undotted appearance of indices referring to B-space. E.g., when σ^{μ}_{Amn} turns up, we replace it by $\sigma^{\mu mn}$, having dropped the subscript B.

If we specialize to the transformations $\exp(i\sigma_k\,\theta_k/2)$ (no sum over k), i.e., exclude all actual Lorentz transformations, then we have in addition to eq. (26a) the condition

$$S_A = S_B \ .$$

See eqs. (8) and (9). These transformations make up the group SU(2), the "special unitary group in two dimensions", special, because the determinant of every one of its matrices is plus 1.

Once we equate in eqs. (22) κ and κ' to m, we have the Dirac equation for the 4-spinor ψ, which is still in the momentum representation:

$$\psi = (1/N)\begin{pmatrix} \xi_{Bs} \\ \xi_{As} \end{pmatrix}, \quad N = 2k\surd(p^0(p^0+m)) \tag{27a}$$

$$(\alpha_k p_k + \beta m)\psi = p^0\psi \tag{27b}$$

where

$$a_k = \begin{pmatrix} \sigma_k & 0 \\ 0 & -\sigma_k \end{pmatrix} \; and \; \beta = \begin{pmatrix} 0 & 1 \\ 1 & 0 \end{pmatrix}. \tag{27c}$$

Due to the derivation presented here, the detailed behavior of the Dirac equation under Lorentz transformation is rather apparent. Including spinor indices, eqs. (23) are

$$-\sigma^{\mu}_{\dot{m}n} p_{\mu} \xi^n_B + m\xi^m_A = 0$$

and (28a)

$$-\sigma^{\mu m\dot{n}} p_{\mu} \xi^n_A + m\xi^m_B = 0 \,.$$

They transform into (written in matrix form), see eqs. (26f and 26g),

$$S^{-1\dagger}\sigma^{\mu} S^{-1} p_{\mu} S\xi_B + mS^{-1\dagger}\xi_A = 0, \quad \sigma^{\mu} = \sigma_k, \sigma^0$$

and (28b)

$$(-S\sigma_k S^\dagger p_k + S\sigma^0 S^\dagger p_0)S^{-1\dagger}\,\xi_A + mS\xi_B = 0$$

respectively. Remember, σ_k without any further specifications is a Pauli sigma, and σ^0 is the unit matrix! We now see that the transformation matrix for the Dirac spinor is given in terms of our 2-spinor transformations by

$$\psi'(p'^\mu) = \mathbf{S}\psi(p^\mu)\,,\ \mathbf{S} = \begin{pmatrix} S & 0 \\ 0 & S^{-1\dagger} \end{pmatrix} . \tag{29}$$

3.1.4 Improper Transformations and Negative Energy Solutions

The Dirac equation is also covariant under "improper transformations". We look at these with the Direc equation in the coordinate representation,

$$(\boldsymbol{\alpha}\bullet\mathbf{p}_{\mathbf{op}} + \beta m)\psi = i\partial\psi/\partial x^0$$

$$\mathbf{p}_{\mathbf{op}} = -i\nabla\,,\ \psi = \psi(\mathbf{p},p^0)\exp(i(\mathbf{p}\bullet\mathbf{r} - p^0x^0)) \tag{30a}$$

$$\psi(\mathbf{p},p^0) \equiv (\psi^m(\mathbf{p},p^0))\,,\ m = 1\ to\ 4$$

1a), space inversion: $\mathbf{r} = -\mathbf{r}'$.

$$\begin{aligned}(\boldsymbol{\alpha}\bullet\mathbf{p}'_{op} + \beta m)\psi' &= i\partial\psi'/\partial x^0 \\ \psi' \equiv \psi'(\mathbf{r}',x^0) &= \eta\beta\psi(\mathbf{r},x^0)\ .\end{aligned} \tag{30b}$$

Eq. (30b) and its solution come about because of the anticommutation

$$\beta\boldsymbol{\alpha} = -\boldsymbol{\alpha}\beta\ . \tag{30c}$$

Inverting twice leads back to the original wave function or its negative, the latter because rotating through 2π multiplies a spinor by minus one (Racah 1937). We thus have

$$\psi'(\mathbf{r},x^0) = \eta\beta\psi(-\mathbf{r},x^0)\,,\ \eta = \pm 1\ \ or\ \pm i \tag{30d}$$

1b), negative energy solutions: If - in the momentum representation - $\psi(\mathbf{p},p^0)$ is a solution, so is $\psi(-\mathbf{p},p^0)$. These spinors can be reached from rest-frame-spinors with components of the same sign (the ξ_u and ξ_d of eqs. (12)) by boosting in opposite directions, or, for that matter, one can be obtained by Lorentz transforming the other. Matters are not so simple as regards $\psi(\mathbf{p},-p^0)$ or $\psi(-\mathbf{p},-p^0)$. Apply the sign change in energy to the spinors of eqs. (27a), and transform to rest: the new spinors vanish unless you also changed the normalization. One can get from m to any positive energy, but a negative energy can only be reached from -m; the invariant interval $p^\mu p_\mu$ is timelike. So let us look at the following equation,

$$(-\boldsymbol{\alpha}\bullet\mathbf{p} - \beta m)\beta\gamma_5\psi(\mathbf{p},p^0) = p^0\beta\gamma_5\psi(\mathbf{p},p^0) \tag{31a}$$

where

$$"\gamma_5" = -i\alpha_1\alpha_2\alpha_3 = \begin{pmatrix} 1 & 0 \\ 0 & -1 \end{pmatrix}\,. \tag{31b}$$

We rewrite it,

$$(\boldsymbol{\alpha}\bullet\mathbf{p} + \beta m)\psi'(\mathbf{p},-p^0) = -p^0\psi'(\mathbf{p},-p^0)\,. \tag{31c}$$

To the solutions for positive energy, see eq. (27a), we can add those of negative energy,

$$\psi_s^{(+)}(p^\mu) = \begin{pmatrix} \xi_{Bs}(p^\mu) \\ \xi_{As}(p^\mu) \end{pmatrix}\,,\ \psi_s^{(-)}(p^\mu) = \begin{pmatrix} -\xi_{As}(p^\mu) \\ \xi_{Bs}(p^\mu) \end{pmatrix};\ s = \pm 1/2\,. \tag{31d}$$

The 4-momentum in $\psi_s^{(-)}(p^\mu)$ is the same as in $\psi_s^{(+)}(p^\mu)$, in particular, p^0 is positive! These are the solutions of eq. (27b) with $\boldsymbol{\alpha}$ and β as in eqs. (27c). - Because the spinors ξ_{As} and ξ_{Bs} satisfy the relativistic Schrödinger or Klein-Gordon equation, see eq. (140a), we know there must also be negative energy solutions. - 2), time reversal (Wigner 1932): $x^0 = -x'^0$

$$\begin{aligned}(\boldsymbol{\alpha} \bullet \mathbf{p}_{\mathbf{op}} + \beta m)\psi' &= i\partial\psi'/\partial x'^0 \\ \psi' \equiv \psi'(\mathbf{r},x'^0) &= \alpha_3\alpha_1\psi^*(\mathbf{r},x^0)\ .\end{aligned} \tag{32a}$$

That ψ' as given is a solution follows from the anti-commutativity of different alpha components and the imaginary character of α_2. A space-time dependence $\exp(i(\mathbf{p}\bullet\mathbf{r} - p^0x^0))$ correctly turns into $\exp(i(\mathbf{p}'\bullet\mathbf{r} - p^0x'^0))$ with $\mathbf{p}'$ the negative of $\mathbf{p}$. The relation

$$\alpha_3\alpha_1 = i\begin{pmatrix}\sigma_2 & 0\\ 0 & \sigma_2\end{pmatrix} \tag{32b}$$

tells us that the matrix multiplying ψ^* is the one which, combined with complex conjugation, reverses the sign of the spin,

$$(i\sigma_2(\boldsymbol{\sigma}/2)(-i\sigma_2))^* = -\boldsymbol{\sigma}/2 \tag{32c}$$

which should happen under time reversal. The standard angular-momentum, $\mathbf{r} \times \mathbf{p}$, classical or quantum, changes sign under time reversal. In the quantum case this requires complex conjugation.

3), charge conjugation: e = -e′

Before discussing this improper transformation, we obviously need to incorporate the electromagnetic interaction in the theory. At this time let us be satisfied with using correspondence, and translate the classical substitution of p^ν by $p^\nu - eA^\nu$ into quantum mechanics as represented by the Dirac equation,

$$(\alpha^k(p_{opk} - eA_k) + \beta m)\psi = (i\partial/\partial x^0 - eA^0)\psi\ . \tag{33a}$$

We render this equation "manifestly covariant",

$$(\gamma^\mu(\partial/\partial x^\mu - ieA_\mu) + m)\psi = 0 \ . \tag{33b}$$

Remember, we are using the metric 1, 1, 1, -1.
Thus

$$\gamma^k = -i\beta\alpha^k, \gamma^0 = -i\beta \ . \tag{33c}$$

The matrix 4-vector γ^u satisfies the anti-commutation relations

$$\{\gamma^\mu,\gamma^\nu\} = 2g^{\mu\nu} \tag{33d}$$

$g^{\mu\nu}$ being the metric tensor. On to charge conjugation. With a given external 4-potential the "charge-conjugate equation" follows from eqs. (33b) to (33d),

$$(\gamma^\mu(\partial/\partial x^\mu - ie'A_\mu) + m)\gamma_2\psi^* = 0 \ . \tag{33e}$$

The particle of mass m and charge -e also obeys the Dirac equation. It is the so-called antiparticle of the original particle. If we let the sources of A_μ participate in the sign change, the interaction obviously does not change: the Dirac equation with an electromagnetic interaction is "charge-conjugation invariant". Going back to the free particle equation, we find that, starting from a positive energy solution $\psi^{(+)}$, γ_2 and complex conjugation produce a negative energy solution with reversed momentum and spin (see also eq. (38b)).

Before we elaborate on the connection between negative energy solutions and charge conjugation, it is indicated to introduce the "conjugate equation" of the Dirac equation. Multiply the Dirac equation by γ^0 and take the γ^0 through to the ψ. Now take the hermitian conjugate. The result is

$$\bar{\psi}(\gamma_\mu(\partial/\partial x^\mu - eA_\mu) - m) = 0 \, , \bar{\psi} = (\gamma^0\psi)^\dagger \ . \tag{34a}$$

γ^0 is a scalar in the sense

$$S^{-1\dagger}\gamma^0 = \gamma^0 S \tag{34b}$$

see eq. (29). Due to this invariance property we can form the scalar density $\bar{\psi}\psi$.

The forms of eqs. (33b) and (33c) exhibit γ^μ as a 4-vector, and the electric 4-current density based on the Dirac equation is

$$j^\mu = e\bar{\psi}\gamma^\mu\psi = e\psi^\dagger(1,\boldsymbol{\alpha})\psi \ . \tag{35a}$$

An explicit demonstration of the proper behavior of j^μ under Lorentz transformations provides a review of the considerations leading up to the Dirac equation. According to eqs. (27) and (28), the 4-vector $(1,\alpha)$ is given by

$$(1,\alpha^k) = -\begin{pmatrix} (\sigma^\mu_{mn}) & 0 \\ 0 & (\sigma^{\mu\dot{m}\dot{n}}) \end{pmatrix} . \tag{35b}$$

We know how ψ transforms, see eqs. (29), and how the sigmas of eq. (35b) transform,

$$S^\dagger \sigma^\mu S = l^\mu{}_\nu \sigma^\nu \, , \, S^{-1}\sigma_\mu S^{-1\dagger} = l^\mu{}_\nu \sigma_\nu \tag{35c}$$

see eqs. (28). The seemingly non-covariant sum is explained by the second one of eqs. (26f), and remembering the space-time metric tensor. Consequently,

$$j'^\mu = l^\mu{}_\nu j^\nu \ . \tag{35d}$$

In relativistic quantum mechanics the pertinent interaction Hamiltonian is $-e\gamma^\mu A_\mu$, see eq. (33b), and using perturbation theory and free particle wave functions we can write down the integrand of a transition amplitude, $-e\bar{\psi}_f\gamma^\mu\psi_i A_\mu$. Now, with positive energies we know how to proceed, due to the lessons of non-relativistic quantum mechanics. When ψ_i and/or ψ_f depend on negative energies, we use the language of quantum field theory, and reinterpret the "destruction" of a negative energy particle as the "creation" of a positive energy antiparticle, and the creation of a negative energy particle as the destruction of a positive energy antiparticle, momentum and spin being reversed in both cases. This kind of interpretation is not particular to Dirac particles (Bogoliubov 1959). As an example we use the above amplitude with ψ_f and ψ_i positive energy electron-functions, and check if it is indeed equal to the corresponding positron amplitude. The functions of the first amplitude are

$$\psi_i = \psi_s^{(+)}(\mathbf{p},p^0)\exp(i(\mathbf{p}\bullet\mathbf{r} - p^0x^0)) \tag{36a}$$

and ψ_f the same, except with primed 4-momentum and spin.

Since negative energy electron-functions are equally good solutions of the Dirac equations, we should also have an amplitude with the function

$$\psi_i = \psi_s^{(-)}(\mathbf{p},p^0)\exp(i(\mathbf{p}\bullet\mathbf{r} + p^0x^0)) \tag{36b}$$

and the same for ψ_f , but again with in general different 4-momentum and arbitrary spin. Now we know that, except for a phase, (a standing for antiparticle)

$$\psi_s^{(-)}(\mathbf{p},p^0) = \gamma^2(\psi_{a-s}^{(+)}(-\mathbf{p},p^0))^* \tag{36c}$$

- see eq. (33e) and subsequent comments as well as eq. (31c) - and so the current density, apart from the exponential factors, becomes

$$j^\mu = -e\psi_{-f}^{(+)T}\gamma^2\gamma^o\gamma^\mu\gamma^2\psi_{-i}^{(+)*} \tag{36d}$$

γ^0 is anti-hermitian. Bringing the two γ^2 together and transposing j^μ results in

$$j^\mu = e\overline{\psi}_{a-i}^{(+)}\gamma^\mu\psi_{a-f}^{(+)}\,. \tag{36e}$$

The minus signs are meant to imply the negatives of the spins and momenta of the negative energy electron-functions. Since spins (along the third axis) and momenta can be picked, the positron current obviously has the same structure as the electron current. - In the quantum-field-theoretical treatment the transposition results in a sign change. ψ and $\bar{\psi}$ are field operators and anticommute.

For the spin dependent relative phase between the functions in eq. (36c) look below.

Contrast this with the situation in weak interactions. Consider electrons and neutrinos. They form a "charge-exchange current" interaction with W particles (Salam 1968, Weinberg 1967), which is

$$j^\mu = g\bar{\psi}_e\gamma^\mu(1-\gamma_5)\psi_\nu \ . \tag{37}$$

Clearly, it is not invariant under charge conjugation, γ^2 and γ^5 do not commute. But check for invariance under "CP", the combined charge conjugation and space inversion. More on this in Section 6.

We end this section with a general expression for the free electron-function,

$$\begin{aligned}\psi_e(x^\mu) = 1/(2\pi)^{-3/2} \int d^3p[a^{(+)}_{es}(\mathbf{p})\, \psi^{(+)}_{es}(\mathbf{p}) \exp(ip^\mu x_\mu) \\ +\eta(s)\, a^{(+)*}_{-e\,-s}(\mathbf{p})\, \psi^{(-)}_{es}(-\mathbf{p}) \exp(-ip^\mu x_\mu)] \ .\end{aligned} \tag{38a}$$

As before, the raised signs in parentheses give the sign of the energy. The interpretation of a destroyed particle of negative energy, spin s and momentum **-p** as a created antiparticle of positive energy, spin -s and momentum **p** is reflected in the substitution of $a^{(-)}_{es}(-\mathbf{p})$ by $a^{(+)*}_{-e-s}(\mathbf{p})$. We sum over the spins, and, for proper behavior under charge conjugation, associate a phase $\eta(s)$ of -1 when s equals +1/2: when operating on the upper and lower components, respectively, of a 4-spinor.

$$\psi = \begin{pmatrix} -\xi_{As} \\ \xi_{Bs} \end{pmatrix}$$

the metric spinors τ_A and τ_B give

$$\psi = \pm \begin{pmatrix} \xi_{B-s} \\ \xi_{A-s} \end{pmatrix}^* , \quad s = \pm 1/2 \ .$$

In the representation of eqs. (12), (27), and (31b) this agrees with

$$-i\beta\gamma_5\,\sigma_2\,\psi_s^{(+)*}(\mathbf{p}) = (-1)^{1/2+s}\,\psi_{-s}^{(-)}(-\mathbf{p}) \ . \tag{38b}$$

3.1.5 Covariants

If we transform ψ into $\boldsymbol{S}\psi$, then $\bar{\psi}$ goes into $\bar{\psi}\boldsymbol{S}^{-1}$. Below we enumerate the transformation properties of various products of Dirac matrices under the similarity transformation with $\boldsymbol{S}$.

1), the scalar: the four by four unit matrix.

2), the 4-vector: according to our convention

$$\gamma^\mu = (-i\beta,\ -i\beta\boldsymbol{\alpha})$$

see eqs. (33c). From eqs. (27c), (29), (35b) and (35c) we obtain

$$\boldsymbol{S}^{-1}\gamma^\mu \mathbf{S} = i\begin{pmatrix} 0 & S^{-1}(\sigma^{\mu m\dot{n}})S^{-1\dagger} \\ S^\dagger(\sigma^\mu_{\dot{m}n})S & 0 \end{pmatrix} = l^\mu{}_\nu\gamma^\nu \ . \tag{39}$$

3), the antisymmetric tensor: the components

$$\sigma^{\mu\nu} = (1/2i)[\gamma^\mu, \gamma^\nu]$$

of which six are independent. E.g.,

$$\gamma_1\gamma_2 = i\sigma_3 \equiv i\sigma_{12}\,,\, \gamma^0\gamma^1 = -\alpha_1 \equiv i\sigma^{01}\ .$$

The Dirac $\sigma^{\mu\nu}$ are obviously the generators of rotations and Lorentz transformations.

4), the pseudoscalar: the matrix $\beta\gamma^1\gamma^2\gamma^3$ called γ_5, see eq. (31a). From item 2) we know that $S^{-1}\gamma^\mu S$ equals $l^\mu{}_\nu\gamma^\nu$. Apply this to $i\gamma^0\gamma^1\gamma^2\gamma^3$ to obtain $||\, l^\mu{}_\nu \,||\, \gamma_5$. For space inversions the determinant equals -1.

5), the axial vector: the four components $i\gamma^\mu\gamma_5$. That the designation is properly chosen follows immediately from items 2) and 4). Items 1) through 5) account for all 16 elements of the "Dirac algebra" (Corson 1953).

We notice that γ^0 plays a dual role. If provided subscripts *m* and *n* it is an invariant second rank spinor. (It serves as a kind of metric spinor in four dimensions.) It is a component of a matrix 4-vector, if transformed by a similarity transformation, $S^{-1}\gamma^0 S$. We encountered a similar situation in two dimensions with σ_2. With subscripts *m* and *n* it is invariant, fulfilling the role of metric spinor. In this role it is commonly multiplied by plus or minus i. With subscripts *m* and *n* or superscripts *m* and *n*, σ_2 is a component of either one of two 4-vectors, see eqs. (28). And there is a third possibility: σ_2 may carry a spinor superscript and subscript, since it may also serve as the component σ_{31} of either one of two antisymmetric tensors $\sigma_{\mu\nu}$,

$$\sigma_{kl} \equiv \sigma_m\,,\, k,l,m = cycl.\,1,2,3\,;\, \sigma_{k0} = \pm i\,\sigma_k\ .$$

See Section 2.

3.1.6 Large and Small Components

A similarity transformation can take us from the α_k and β of eqs. (24c) to

$$\alpha_k = \begin{pmatrix} 0 & \sigma_k \\ \sigma_k & 0 \end{pmatrix}\,,\; \beta = \begin{pmatrix} 1 & 0 \\ 0 & -1 \end{pmatrix}\ . \tag{40}$$

The solutions of the free Dirac equation going with these matrices are

$$\psi^{(+)}_{1/2}(\mathbf{p},p^0) = \frac{1}{N}\begin{pmatrix} 1 \\ 0 \\ p_3/(p^0+m) \\ (p_1+ip_2)/(p^0+m) \end{pmatrix}, \psi^{(+)}_{-1/2}(\mathbf{p},p^0) = \frac{1}{N}\begin{pmatrix} 0 \\ 1 \\ (p_1-ip_2)/(p^0+m) \\ -p_3/(p^0+m) \end{pmatrix}$$

$$\psi^{(-)}_{1/2}(\mathbf{p},p^0) = \frac{1}{N}\begin{pmatrix} -p_3/(p^0+m) \\ -(p_1+ip_2)/(p^0+m) \\ 1 \\ 0 \end{pmatrix}, \psi^{(-)}_{-1/2}(\mathbf{p},p^0) = \frac{1}{N}\begin{pmatrix} -(p_1-ip_2)/(p^0+m) \\ p_3/(p^0+m) \\ 0 \\ 1 \end{pmatrix}.$$

These 4-spinors are particularly useful when the departure from the non-relativistic domain is not considerable. We then speak of "large and small (4-spinor) components". As with the previously written solutions, the **p** and s refer to the initial or “destroyed” quantities. The plane wave factors are $\exp(i(\mathbf{p}\bullet\mathbf{r} \mp p^0x^0))$ for the $\psi^{(+)}$ and $\psi^{(-)}$, respectively. (Obviously, the upper sign goes with the $\psi^{(+)}$ spinors). If we normalize to, say, one electronic charge, i.e., to $\psi^\dagger\psi = 1$, the normalization coefficient is $\sqrt{((p^0+m)/2p^0)}$. For the invariant normalization, $\bar{\psi}\psi = 1$, it is $\sqrt{((p^0+m)/2m)}$, see eq. (34a).

3.1.7 Longitudinal Polarization

Weak interactions give rise to longitudinally polarized particles (Frauenfelder 1957). Eq. (37) describes the weakly interacting electron-neutrino-current. $(1-\gamma_5)\psi_\nu$ describes, as we know, not only the destruction of a (positive energy) neutrino, which is left handed, but also the creation of an antineutrino, which is right handed. In eq. (37) we can take the $(1-\gamma_5)$ through the Dirac matrices and place it in front of the electron function. This describes the creation of a “left-handed” electron. Let us calculate the expectation value

$$<\boldsymbol{\sigma}\bullet\mathbf{p}>/p = \sum_s \xi^\dagger_{As}\,\boldsymbol{\sigma}\bullet\mathbf{p}\,\xi_{As}/N^2p \tag{41a}$$

see eq. (27a): we have

$$\boldsymbol{\sigma}\bullet\mathbf{p}\,\xi_{As} = -p^{0}\xi_{As} + m\xi_{Bs}$$

from eq. (23b). So we calculate the products $\xi^{\dagger}_{As}\xi_{Bs}$ and $\xi^{\dagger}_{As}\xi_{Bs}$ and sum over s; see eqs. (12). We obtain

$$\langle\boldsymbol{\sigma}\bullet\mathbf{p}\rangle/p = -p/p^{0} = -v \ \ in\ units\ of\ c\ . \tag{41b}$$

With no mass, so called handedness and "helicity" are the same [A.5].

3.2 MASSLESS PARTICLES

A 2-spinor can quite generally be written

$$\chi = a\begin{pmatrix}1\\0\end{pmatrix} + b\begin{pmatrix}0\\1\end{pmatrix}\ . \tag{42a}$$

Normalization lets us equate

$$a = \cos\theta/2,\ \ b = \sin\theta/2\ \exp(i\varphi)\ .$$

We now wish to describe massless spin-one-half particles.

In our discussion of 2-spinors as spin functions of massive particles the three-dimensional rotation group figured prominently; see section 3.1.1. This was so because the 4-vector state

$$V_k = 0,\ V^0 \neq 0 \tag{43}$$

remains invariant under rotations, and the spinors, like 4-vectors belonging to representations of the Lorentz group, L_h, were expressed as functions of the 4-momentum.

3.2.1 The Two-Dimensional Euclidian Group

While the up and down spinors change of course under rotations, their simple forms suggest the question, whether they are not invariant under some subgroup of L_h. They remain unaltered when multiplied by

$$\begin{pmatrix} 1 & a \\ 0 & 1 \end{pmatrix} \quad and \quad \begin{pmatrix} 1 & 0 \\ a & 1 \end{pmatrix} \tag{44a}$$

respectively. Let us express these matrices as infinitesimal transformations

$$1 + i\varepsilon \begin{pmatrix} 0 & 1 \\ 0 & 0 \end{pmatrix} \quad and \quad 1 + i\varepsilon \begin{pmatrix} 0 & 0 \\ 1 & 0 \end{pmatrix} \tag{44b}$$

and in turn the generating matrices in terms of Pauli sigmas

$$\begin{pmatrix} 0 & 1 \\ 0 & 0 \end{pmatrix} = (i\sigma_2 + \sigma_1)/2 \quad and \quad \begin{pmatrix} 0 & 0 \\ 1 & 0 \end{pmatrix} = (-i\sigma_2 + \sigma_1)/2 \ . \tag{44c}$$

We remember $i\sigma_2$ as a standard-Lorentz-transformation generator of the representation (0,1/2) and a boost of (1/2,0). See eqs. (9). For a reason which will immediately become apparent, we shall associate the first matrix with (0,1/2) and the second with (1/2,0).

The generating matrix corresponding to the above matrices, which is applicable in the transformations of 4-vectors, is

$$K_2 + J_1 = \begin{pmatrix} 0 & 0 & 0 & 0 \\ 0 & 0 & -i & i \\ 0 & i & 0 & 0 \\ 0 & i & 0 & 0 \end{pmatrix} . \tag{45a}$$

It does not affect the state

$$V_1 = V_2 = 0,\ V^0 = V_3 \ . \tag{45b}$$

J_k and K_k are here the symbols for rotation and Lorentz-transformation generators appropriate to 4-space. The fact that an infinitesimal transformation with the above generating matrix keeps the energy equal to the momentum along the positive third axis explains the choice of association we just made. Also the matrix

$$K_1 - J_2 = \begin{pmatrix} 0 & 0 & -i & i \\ 0 & 0 & 0 & 0 \\ i & 0 & 0 & 0 \\ i & 0 & 0 & 0 \end{pmatrix} \tag{45c}$$

leaves the vector state of eq. (45b) alone, and so does the rotation generator J_3.

The two-by-two matrices corresponding to the matrix of eq. (45c) are

$$\begin{pmatrix} 0 & i \\ 0 & 0 \end{pmatrix} = (i\sigma_1 - \sigma_2)/2 \ \textit{and} \begin{pmatrix} 0 & 0 \\ -i & 0 \end{pmatrix} = -(i\sigma_1 + \sigma_2)/2 \ . \tag{46}$$

Calling the general matrices corresponding to the matrices of eqs. (45c) and (46) and of eqs. (45a) and (44c) E_1 and E_2, respectively, we find the commutation relations.

$$[E_1,E_2] = 0,\ [E_2,J_3] = iE_1 \ \textit{and} \ [J_3,E_1] = iE_2 \ . \tag{47}$$

- Remember, the rotation generators are the same for (1/2,0) and (0,1/2). - Eqs. (47) define the "two-dimensional Euclidian group, E(2)," since E_1 and E_2 correspond to translational generators along Cartesian axes one and two, and J_3 to the rotation generator in the one-two plane. The finite two-dimensional group elements are easily evaluated as

$$E_B(\eta_1,\eta_2,\varphi) = \begin{pmatrix} \exp(i\varphi/2) & (i\eta_2-\eta_1)\exp(\pm i\varphi/2) \\ 0 & \exp(-i\varphi/2) \end{pmatrix} \quad and \tag{48a}$$

$$E_A(\eta_1,\eta_2,\varphi) = \begin{pmatrix} \exp(i\varphi/2) & 0 \\ (i\eta_2+\eta_1)\exp(\mp i\varphi/2) & \exp(-i\varphi/2) \end{pmatrix} \tag{48b}$$

η_1 and η_2 being the parameters going with E_1 and E_2, respectively. There is a choice of signs because of the order in which the rotation and the "translations" are applied. Rotating last, the corresponding four-by-four matrix is

$$E(\eta_1,\eta_2,\varphi) = \begin{pmatrix} \cos\varphi & \sin\varphi & \eta_1\cos\varphi+\eta_2\sin\varphi & -\eta_1\cos\varphi-\eta_2\sin\varphi \\ -\sin\varphi & \cos\varphi & -\eta_1\sin\varphi+\eta_2\cos\varphi & \eta_1\sin\varphi-\eta_2\cos\varphi \\ -\eta_1 & -\eta_2 & 1-(\eta_1^2+\eta_2^2)/2 & (\eta_1^2+\eta_2^2)/2 \\ -\eta_1 & -\eta_2 & -(\eta_1^2+\eta_2^2)/2 & 1+(\eta_1^2+\eta_2^2)/2 \end{pmatrix}. \tag{49}$$

Thus if we associate the up-spinor transforming according to (0,1/2) with the up-momentum, then subjecting it to the group E(2), leaving $p_3 > 0$ alone, will merely change its phase. The same conclusion is reached for the down-spinor transforming according to (1/2,0) and also associated with p_{up}.

3.2.2 Wigner Rotations and Helicity

Just as under the heading SPINORS, we generate standard spinors, except we leave out the matrix $R(\theta,\varphi)$. This means, we turn our up and down-spinors into the spinors

$$\chi_{B\,1/2} = \exp(\alpha/2)\begin{pmatrix} \exp(-i\varphi/2)\cos\theta/2 \\ \exp(i\varphi/2)\sin\theta/2 \end{pmatrix} \tag{50a}$$

$$\chi_{A\,-1/2} = \exp(\alpha/2)\begin{pmatrix} -\exp(-i\varphi/2)\sin\theta/2 \\ \exp(i\varphi/2)\cos\theta/2 \end{pmatrix}$$

Introducing explicit momentum dependence with m = 0, changing overall phases, and normalizing with

$$N = \surd(2p(p+p_3)) \;\; and \;\; N' = \surd(2p(p-p_3))$$

respectively, we obtain

$$\chi_{1/2} = 1/N\begin{pmatrix} p+p_3 \\ p_1+ip_2 \end{pmatrix} = \exp(i\varphi)/N'\begin{pmatrix} p_1-ip_2 \\ p-p_3 \end{pmatrix}$$

$$\chi_{-1/2} = 1/N\begin{pmatrix} -(p_1-ip_2) \\ p+p_3 \end{pmatrix} = -\exp(-i\varphi)/N'\begin{pmatrix} p-p_3 \\ -(p_1+ip_2) \end{pmatrix} \tag{50b}$$

$$where \qquad p = p_i \exp\alpha \, , p_i = p \; initial \; .$$

The functions χ_s have "helicity", i.e.,

$$(\boldsymbol{\sigma}\bullet\mathbf{p}/p)\chi_s = \pm\chi_s \; . \tag{51}$$

The capital indices were left off the normalized spinors to distinguish them from the unnormalized ones. This does not give rise to an ambiguity, since we have, for given **p**, only two spinors.

The considerations revolving around eqs. (13) and (14) can be repeated, except that the standard spinors are now those of eqs. (50a), and the transformations T_C of eqs. (10) through (12) are replaced by

$$H_C = R^{-1}(\theta,\varphi)B_C(\alpha) \, , \, C = A \; or \; B \; . \tag{52}$$

One arrives again at eq. (14d), the spinors now being the χ_{Cs}. The analogue of the matrix R_W of eq. (14a) then becomes one of the matrices of eqs. (48). But remember, our "massless-basis-spinors" are restricted to $\begin{pmatrix}1\\0\end{pmatrix}$ for B = ½ and $\begin{pmatrix}0\\1\end{pmatrix}$ for A = ½ ! Therefore, the only nonvanishing elements of the Wigner rotation matrix are

$$t'^{r}_{1/2} = \delta^{r}_{1/2} \exp(i\varphi/2) \; for\, B \;\; and \;\; t'^{r}_{-1/2} = \delta^{r}_{-1/2} \exp(-i\varphi/2) \; for\, A \; . \qquad (53)$$

Clearly, the situation is much simpler than for massive particles: except for phase factors, the group E(2) leaves certain spin-one-half states invariant: helicity becomes a particle property.

It is useful to introduce the concept helicity also into the discussion of massive particles. The transformation T_C operating on, say, an up-spinor may trivially be related to the transformation H_C operating on up- and down-spinors,

$$R^{-1}B_C R\begin{pmatrix}1\\0\end{pmatrix} = R^{-1}B_C\begin{pmatrix}1 & 0\\0 & 1\end{pmatrix} R_{1/2} \qquad (54)$$

$R_{1/2}$ being the first column of the rotation matrix, and the unit matrix on the right representing the (ordinary, not direct) sum of up- and down-spinors. Written out, this equation becomes

$$\xi_{C\,1/2} = \exp(i\varphi/2)(\cos\theta/2\,\chi_{C\,1/2} - \sin\theta/2\,\chi_{C\,-1/2}) \qquad (55)$$

with the "helicity functions"

$$\chi_{C\,1/2} = \exp(\pm\alpha/2)\begin{pmatrix}\exp(-i\varphi/2)\cos\theta/2\\ \exp(i\psi/2)\sin\theta/2\end{pmatrix} \; and$$
$$\chi_{C\,-1/2} = \exp(\mp\alpha/2)\begin{pmatrix}-\exp(-i\varphi/2)\sin\theta/2\\ \exp(i\varphi/2)\cos\theta/2\end{pmatrix} \qquad (56)$$

the upper and lower signs for C equaling B and A, respectively. The coefficients of the linear combination $\xi_{C\text{-}1/2}$ should be obvious.

The direct sums

$$\chi_{B\,1/2} \oplus \chi_{A\,1/2} \; and \; \chi_{B\,-1/2} \oplus \chi_{A\,-1/2} \tag{57}$$

are the positive energy Dirac spinors in the helicity representation. At high energy they approach two-component spinors.

3.2.3 The Weyl Equations

The equations

$$S^{m}_{Cn}\,\chi^{n}_{Cs}(p^{\mu}) = \exp(\pm i\,\varphi/2)\,\chi^{m}_{Cs}(p^{\mu}) \tag{58}$$

with S^{m}_{Cn} an arbitrary Lorentz transformation in (0,1/2) or (1/2,0) and χ^{n}_{Cs} given by eqs. (50a), follows from our discussions in sections 3.2.1 and 3.2.2, eqs. (52) and (53) being of particular relevance. Thus helicity goes with masslessness, and is not just a property of a particular state, as it is with massive particles. This can also be shown by checking the covariance (in the general sense) of the eqs. (51) with the solutions given by eqs. (50a) or (50b). These equations can also be written

$$\boldsymbol{\sigma}\bullet\mathbf{p}\,\chi_{s}(x^{\mu}) = \pm p^{0}\chi_{s}(x^{\mu})\,,\;\; s = \pm 1/2 \tag{59a}$$

in the coordinate representation. They are the Weyl equations (Weyl 1929). In the momentum representation they can, as homogeneous equations, only depend on two coordinates (or parameters). The solutions (50a) make that quite clear; the boost can always be normalized away, the big difference from the massive case. There we must accommodate a third parameter. Consequently, we get inhomogeneous equations; see eqs. (23). Combining the inhomogeneous equations, we get a homogeneous 4-spinor-equation, the Dirac equation; see eqs. (27). But then three more parameters may appear. Since massive particles do not have helicity, there are, for spin-one-half, two spin states to superpose. This allows us to choose the direction of the spin independently of the momentum, using up two parameters. Applied to neutrinos, the general solutions of the first of eqs. (59a) is (the subscript $\bar{\nu}$ means "antineutrino"; s = 1/2)

$$\chi_{\bar{\nu}}(x^{\mu}) = (2\pi)^{-3/2}\int d^3p[a_{\bar{\nu}}^{(+)}(\mathbf{p})\chi_{\bar{\nu}}^{(+)}(\mathbf{p})\exp(ip^{\mu}x_{\mu}) + a_{\bar{\nu}}^{(-)}(-\mathbf{p})\chi_{\bar{\nu}}^{(-)}(-\mathbf{p})\exp(-ip^{\mu}x_{\mu})] \tag{59b}$$

where, see eqs. (50b),

$$\chi_{\bar{\nu}}^{(+)}(\mathbf{p}) = \chi_{1/2}$$

and $\chi_{\bar{\nu}}^{(-)}(-\mathbf{p})$ obviously the same (see eq. (59a)) except for a phase factor, which we choose to be plus one. From a general solution of one Weyl equation one may get one of the other:

$$\chi_{\nu}(x^{\mu}) = \tau\chi_{\bar{\nu}}^{*}(x^{\mu}) \ . \tag{59c}$$

Since the spinors $\chi_{-1/2}$ and $\chi_{1/2}$ are so related, see eqs. (50b), we can replace the $\chi_{\bar{\nu}}^{(-)}(-\mathbf{p})$ in eq. (59b) by $-\tau\chi_{\nu}^{(+)*}(\mathbf{p})$,

$$\chi_{\nu}^{(+)}(\mathbf{p}) = \chi_{\nu}^{(-)}(-\mathbf{p}) = \chi_{-1/2} \ .$$

Then, if we wish the general solutions to satisfy eq. (59c), we obtain

$$a_{\bar{\nu}}^{(-)}(-\mathbf{p}) = a_{\nu}^{(+)*}(\mathbf{p}) \ \textit{and}\ a_{\nu}^{(-)}(-\mathbf{p}) = a_{\bar{\nu}}^{(+)*}(\mathbf{p}) \ . \tag{59d}$$

- In quantum field theory one says, as we can deduce from the example of Dirac electrons and positrons, that the destruction of a negative energy antineutrino (neutrino) is in effect the creation of a positive energy neutrino (antineutrino) of opposite momentum. Note that electron and positron obey the same equation, except for the sign of the charge (when the electromagnetic interaction is taken into account); Weyl neutrino and antineutrino obey different equations. -

3.3 SPINOR REPRESENTATION OF THE MOMENTUM

In section 3.1.2 we established a formulation of 4-vectors as second rank spinors. With the spinors of eqs. (50a) and (56) it is now easy to obtain explicit relations. Consider the combination

$$\chi_{A\,-1/2}^{1/2}\,\chi_{B\,1/2}^{-1/2} - \chi_{A\,1/2}^{1/2}\,\chi_{B\,-1/2}^{-1/2} = -\exp(\alpha)(1-\cos\theta)/2 - \exp(-\alpha)(1+\cos\theta)/2 = (p_3+p_0)/m\,. \quad (60a)$$

Similarly we obtain

$$\chi_{A\,-1/2}^{-1/2}\,\chi_{B\,1/2}^{1/2} - \chi_{A\,1/2}^{-1/2}\,\chi_{B\,-1/2}^{1/2} = (p_3-p_0)/m\ . \quad (60b)$$

Also,

$$\chi_{A\,-1/2}^{1/2}\,\chi_{B\,1/2}^{1/2} - \chi_{A\,1/2}^{1/2}\,\chi_{B\,-1/2}^{1/2} = (-\exp(\alpha)+\exp(-\alpha))\exp(-i\varphi)(\sin\theta)/2 = -(p_1-ip_2)/m \quad (60c)$$

and similarly

$$\chi_{A\,-1/2}^{-1/2}\,\chi_{B\,1/2}^{-1/2} - \chi_{A\,1/2}^{-1/2}\,\chi_{B\,-1/2}^{-1/2} = (p_1+ip_2)/m\ . \quad (60d)$$

For zero mass the second products on the left all vanish and the right sides turn out to be

$$p_3+p_0,\, p_3-p_0,\, -(p_1-ip_2)\ and\ p_1+ip_2 \quad (60e)$$

in units of $2p_i$, p_i being the initial momentum, usually set equal to one.

In the massive case the exchange

$$\chi_{As}^{m} \rightleftarrows \chi_{Bs}^{m}$$

within the 4-spinors effects a space inversion, **p** goes into **-p** (Weyl 1931). Compare this with the pertinent remarks in section 3.1.4. In the massless case the usual interpretation is to consider $\chi_{1/2}$ and $\chi_{-1/2}$ separately as one another's antiparticle spinors. So there is no space-inversion invariance. See eqs. (50b) and (59c).

3.4 THE COVERING GROUP

In the course of discussing the behavior of the Dirac equation under space inversion, see section 3.1.4, we encountered the doublevaluedness of spinor transformations. Now we look at this property in some detail. Eq. (21c) is a good starting point. It obviously is invariant under multiplication of the spinor-transformation-matrices by minus one. A general spinor transformation, S, involves an actual Lorentz transformation along some axis followed by a three-dimensional rotation:

$$S = R(\theta,\varphi,\psi)\, B(\alpha)\, R(\theta',\varphi') \ . \tag{61}$$

We may include a third primed angle. The R with three angles is conveniently obtained by multiplying the standard R, see eq. (11a), from the left by a rotation about the third axis. Instead of using such an R, now with a plus and minus sign (!), we may use

$$R = \exp(i\boldsymbol{\sigma}\bullet\hat{\mathbf{r}}\,\psi/2)\,,\ O \le \psi \le 2\pi \ . \tag{62a}$$

The unit vector has components sinθ cosφ, sinθ sinφ and cosθ, where

$$0 \le \theta \le \pi \ \mathit{and}\ 0 \le \varphi \le 2\pi \ .$$

The boost is plus and minus that of eq. (11b).

The three-by-three rotation matrix corresponding to expression (62a) in a one to two manner may be written in the analogous form

$$R = \exp(i\,\mathbf{J}\cdot\hat{\mathbf{r}}\,\psi)\,,\ 0 \le \psi \le \pi \ . \tag{62b}$$

- While all rotations through 2π give the same two-by-two matrix, in the three-by-three case only the rotations about oppositely directed unit vectors $\hat{r}$ and $\hat{r}'$ through π give the same result (Wigner 1931).- Eq. (62a) may also be written

$$R = \cos\ \psi/2 + i\,\boldsymbol{\sigma}\cdot\hat{\mathbf{r}}\ \sin\ \psi/2 \ .$$

Evidently, rotating through 2π gives minus one. As observed before, see eq.

(24b), the spinor transformations have determinant +1. "The spinor transformations define the two-dimensional unimodular group," and as elements of this group the matrices form a single-valued representation. This group is also called "SL(2,c)", the special linear group in two dimensions with complex coefficients. It is the "covering group" of L_h. All of its representations are single valued, and none of its subgroups are "homomorphic unto L_h," i.e., none are multi-valued representations of L_h (Fonda 1970).

3.5 FOUR-COMPONENT NEUTRINOS AND GAUGE TRANSFORMATIONS

In section 3.2.2 helicity was discussed, as a particle property, which it is for massless particles, and as merely the property of particular states. Thus the expressions (57) are the positive-energy Dirac spinors for a massive particle in the two helicity states. If we let m go to zero, so do $\chi_{A\,½}$ and $\chi_{B\text{-}1/2}$. Consequently, two components of each of the 4-spinors become zero. Remember,

$$\exp(-a) = m/(p^0+p) \rightarrow 0 \;\; as\; m \rightarrow 0 \; .$$

Also, the Dirac equation reduces to the two Weyl equations; see eqs. (27). If, on the other hand, we start with a massless particle, in which case

$$\exp(-\alpha) = p_i/p, \quad p_i = p\; initial$$

we need to remember: when the momentum "points up" along the third axis, so does the basis spinor of the representation (0, 1/2), while the basis spinor of (1/2, 0) points down; see section 3.2.1. In this case $\chi_{A\,½}$ and $\chi_{B\text{-}1/2}$ are automatically zero, no limiting process being required. Eq. (55) still gives us the + ½ two-component spinors making up a Dirac spinor, but not in the helicity representation. Even for massless particles the Dirac spinors then has four nonvanishing components.

The general expression for the Dirac-neutrino-function is obtained from that for the electron, see eq. (38), by replacing the subscripts e and -e by ν and $\bar{\nu}$. The wave functions may, of course, be taken in any representation. For $m = 0$ the representation with the 4-spinors

$$\psi^{(+)}_{\nu\,-1/2} = 1/N \begin{pmatrix} 0 \\ 0 \\ -(p_1 - ip_2) \\ p + p_3 \end{pmatrix} \quad \psi^{(+)}_{\nu\,1/2} = 1/N \begin{pmatrix} p + p_3 \\ p_1 + ip_2 \\ 0 \\ 0 \end{pmatrix} \tag{63}$$

see eqs. (50b), is the most natural. The 4-spinors of eqs. (40) (for massless neutrinos set m = 0), sometimes with the two top and two bottom components and/or the positive and negative energy states interchanged, appear rather frequently in the literature.

The Fourier integral of eq. (38) adapted to the Dirac neutrino (see above) is easily changed to one representing the "Majorana Neutrino" (Majorana 1937). All one need do is replace the subscripts ν by ν. The integrand of $\psi_M(x^\mu)$, M for "Majorana", then is

$$a^{(+)}_{\nu s}(\mathbf{p})\, \psi^{(+)}_{\nu s}(\mathbf{p}) \exp(ip^\mu x_\mu) + a^{(+)*}_{\nu s}(\mathbf{p})\, \gamma_2\, \psi^{(+)*}_{\nu s}(\mathbf{p}) \exp(-ip^\mu x_\mu) \tag{64}$$

s summed from -1/2 to +1/2: the Majorana neutrino is "self-charge-conjugate,"

$$\psi_M(x^\mu) = C\psi^*_M(x^\mu)\,,\ C = \gamma_2 = -i\beta\alpha_2\,. \tag{65}$$

Using the $\psi^{(+)}_{\nu s}$ of eqs. (63), we immediately see an equivalence of the (massless) Majorana neutrino and the Weyl neutrinos (Serpe 1952). Thus the Weyl antineutrino function, $\chi_{\bar\nu}(\mathbf{p})$, is made up of the upper components of $\psi^{(+)}_{\nu\,1/2}$. However, the Weyl neutrinos permit a "gauge transformation of the first kind"

$$\chi_{\bar\nu}(x^\mu) \to \chi_{\bar\nu}(x^\mu) \exp(-i\alpha),\ \chi_\nu(x^\mu) \to \chi_\nu(x^\mu) \exp(i\alpha) \tag{66}$$

the Majorana neutrino does not (see eq. (65)). Although electrically neutral, the Weyl neutrinos may carry other charges, the Majorana neutrino may not (see Section 6). Looking at the electron-neutrino-current, eq. (37), we see that due to the $1\text{-}\gamma_5$ only the left-handed part of the neutrino is participating, giving the appearance that a gauge transformation can always be applied. - With the Weyl neutrino the $1-\gamma_5$ is automatically taken care of, it need not appear in the current. - If a $1+\gamma_5$ admixture occurs, and the neutrino is a Majorana neutrino, there is manifestly no gauge invariance of the current. With a Dirac neutrino the current can be made gauge invariant, admixture or no admixture. One simply lets electron and positron gauge transform like the Weyl neutrinos in eq. (66). But in that case discovering an admixture would be very difficult, because, differently from the Majorana neutrino, a Dirac neutrino cannot be reabsorbed by the same current which caused its emission. That would violate gauge invariance. Consequently, a Dirac neutrino cannot make "neutrinoless double-beta decay" happen, the most conspicuous hypothetical process distinguishing between Dirac and Majorana neutrinos. See Section 7 for a detailed treatment.

While the Weyl neutrinos are massless, the Majorana neutrino, just like the Dirac neutrinos, need not be. A massive $\psi_M(x^\mu)$ satisfying the condition (65) can obviously be constructed.

3.6 SOME COMPARISONS WITH EXPERIMENT

Before the realization that parity is not conserved in weak interactions (Lee 1956), and the subsequent formulation of interactions with neutrinos in terms of Weyl neutrinos (Lee 1957, Landau 1957, Salam 1957), it was standard practice to use Dirac neutrinos in the calculation of decay processes and, to the small extent that this was done, in the calculation of neutrino induced reactions. Evaluations of the weak coupling-constant were based on the comparison of calculations and experiments involving weak-decay-rates. Using resulting values in calculations of neutrino induced reaction cross-sections permitted establishing the feasability of neutrino experiments and then the comparison of theory with experiment.

In calculating the total cross-section of the first neutrino induced process investigated experimentally (Cowan 1956, Reines 1959),

$$\nu_a + p \rightarrow n + e^+$$

one needs to average over the spin states of the incident system. The proton is equally distributed over its two spin states, and so right handed and left handed Dirac antineutrinos would encounter the same cross section. The decay

$$n \rightarrow p + e^- + \nu_a$$

proceeds at a rate which, for given coupling strength, would with two neutrino spin states be twice as fast as with one (one sums over final states). Therefore, a cross-section calculation using a coupling constant based on the evaluation of a decay rate will with one neutrino state yield twice the cross section that it does with two states (Muehlhause 1957, King 1958 (1)).

The earliest data on the "inverse beta-decay of the neutron" were in better agreement with two antineutrino states than with one (King 1958 (2)). The experimenters were attempting to measure a cross section of an order of magnitude of 10^{-43} cm^2 ! The "Cowan-Reines Experiments" involved reactor antineutrinos and water plus cadmium chloride as target. The idea was that an emitted positron quickly forms positronium, which annihilates into gammas, while the associated neutron takes microseconds to slow down, before being captured with the emission of gammas. Two light pulses in delayed coincidence, originating in a liquid scintillator, signal an event. -Once the nonconservation of parity in weak interactions was established, and maximally at that (Wu 1957, Postma 1957), a quick succession of incisive experiments pinned down the (low energy) form of the (to be precise: "semi leptonic") weak interaction, including the helicity of the neutrino (Jackson 1958).

One of the most beautiful physics experiments ever performed established the helicity of the neutrino, defined as the particle associated with the electron in the weak interaction, see eq. (37). The experiment involved K-capture in Eu^{152m} and deexcitation by gamma emission with subsequent resonance scattering by Sm^{152} and counting in NaI, deducing the sought after helicity from the circular polarization of the gamma rays. Before scattering, the gamma rays had to traverse magnetized iron to allow the determination of their polarization. The neutrinos emitted oppositely to the counted gammas were deduced to have negative helicity (Goldhaber 1958). In eq. (37) we have a $1-\gamma_5$, in agreement with having in the beginning placed the two-component spinors $\xi_{B\,s}$ in the upper spots of the Dirac spinor, and the $\xi_{A\,s}$ in the lower spots; see eqs. (12) and (27).

Chapter 4

VECTORS

The representations (1,0) and (0,1) are connected with generally more familiar matters than are the representations (1/2,0) and (0,1/2). Since they are three-dimensional, the three-dimensional rotation group, the most familiar subgroup of L_h, appears in the "regular representation". Choosing real orthogonal matrices for this representation implies a real 3-vector subspace, of which the Cartesian coordinates are an example. Paraphrasing, the J_k can be made up of matrix elements J^l_{km} which transform in the same way with respect to all three indices. Furthermore, as we shall see, second-rank antisymmetric tensors belong to the representations (1,0) and (0,1), e.g., the electromagnetic field. We start with the matrices J_k

$$J_1 = \begin{pmatrix} 0 & 0 & 0 \\ 0 & 0 & -i \\ 0 & i & 0 \end{pmatrix}, \ J_2 = \begin{pmatrix} 0 & 0 & i \\ 0 & 0 & 0 \\ -i & 0 & 0 \end{pmatrix}, \ J_3 = \begin{pmatrix} 0 & -i & 0 \\ i & 0 & 0 \\ 0 & 0 & 0 \end{pmatrix} \tag{67}$$

which generate rotations about x_k according to the well known recipe

$$x'_l = r_{lm} x_m, \ r_{lm} = r^l{}_m = \lim_{n \to \infty} (1 + i\theta J^l_{km}/n)^n . \tag{68}$$

So as to be able to use vectors as simple as the up and down spinors, we subject the J_k to a similarity transformation $U'^{-1} J_k U'$ with the matrix whose columns contain the components along the coordinates x_k of the three eigenvectors of the rotations about the third axis [A.4],

$$U' = (1/\sqrt{2}) \begin{pmatrix} -1 & 0 & 1 \\ -i & 0 & -i \\ 0 & \sqrt{2} & 0 \end{pmatrix} . \tag{69}$$

We obtain the well known angular momentum matrices

$$J_1 = (1/\sqrt{2})\begin{pmatrix} 0 & 1 & 0 \\ 1 & 0 & 1 \\ 0 & 1 & 0 \end{pmatrix},\; J_2 = (1/\sqrt{2})\begin{pmatrix} 0 & -i & 0 \\ i & 0 & -i \\ 0 & i & 0 \end{pmatrix},\; J_3 = \begin{pmatrix} 1 & 0 & 0 \\ 0 & 0 & 0 \\ 0 & 0 & -1 \end{pmatrix}. \tag{70}$$

We can now produce standard general vectors with three-by-three transformations corresponding to the two-by-two transformations of eqs. (10) through (12), or with three-by-three helicity preserving transformations, where, like in the two-by-two case, we first boost along the third axis and then rotate the coordinates; see eq. (52). The basis vectors these matrices are applied to are

$$v_u = \begin{pmatrix} 1 \\ 0 \\ 0 \end{pmatrix},\; v_o = \begin{pmatrix} 0 \\ 1 \\ 0 \end{pmatrix},\; v_d = \begin{pmatrix} 0 \\ 0 \\ 1 \end{pmatrix}. \tag{71}$$

The rotation matrices are of course obtained from the corresponding real orthogonal ones, see eqs. (3a) and their cyclic permutations, by a similarity transformation with the same matrix U'. The boost along the third axis is rather obviously

$$B_C = \begin{pmatrix} \exp(\mp\alpha) & 0 & 0 \\ 0 & 0 & 0 \\ 0 & 0 & \exp(\pm\alpha) \end{pmatrix},\; C = A \, or \, B\,. \tag{72}$$

4.1 THE EUCLIDIAN GROUP AND HELICITY

For massive particles all vectors in eqs. (71) are necessary to generate a set of standard general vectors belonging to (1,0) and (0,1). For massless particles we again get helicities independent of the observer by associating one spin state with one momentum in the case of one representation, and the opposite spin state with the same momentum as before for the other representation. The basis vector with spin zero along the third axis does not enter. We can see all this by looking, as before, at the Euclidian subgroup, E(2). The generators E_1 and E_2 are now

$$E_1 = \sqrt{2}\begin{pmatrix} 0 & i & 0 \\ 0 & 0 & i \\ 0 & 0 & 0 \end{pmatrix}, E_2 = \sqrt{2}\begin{pmatrix} 0 & 1 & 0 \\ 0 & 0 & 1 \\ 0 & 0 & 0 \end{pmatrix} \quad \textit{for } B \neq 0 \tag{73a}$$

$$E_1 = \sqrt{2}\begin{pmatrix} 0 & 0 & 0 \\ -i & 0 & 0 \\ 0 & -i & 0 \end{pmatrix}, E_2 = \sqrt{2}\begin{pmatrix} 0 & 0 & 0 \\ 1 & 0 & 0 \\ 0 & 1 & 0 \end{pmatrix} \quad \textit{for } A \neq 0 \tag{73b}$$

and the general group elements

$$E_B(\eta_1, \eta_2, \varphi) = \begin{pmatrix} 1 & -\sqrt{2}(\eta_1 - i\eta_2) & (\eta_1 - i\eta_2)^2 \\ 0 & 1 & -\sqrt{2}(\eta_1 - i\eta_2) \\ 0 & 0 & 1 \end{pmatrix} \quad \textit{for } B \neq 0 \tag{74a}$$

$$E_A(\eta_1, \eta_2, \varphi) = \begin{pmatrix} 1 & 0 & 0 \\ \sqrt{2}(\eta_1 + i\eta_2) & 1 & 0 \\ (\eta_1 + i\eta_2)^2 & \sqrt{2}(\eta_1 + i\eta_2) & 1 \end{pmatrix} \quad \textit{for } A \neq 0 \tag{74b}$$

multiplied from the left or right by the matrix for the rotation about the third axis; compare to eqs. (48). Clearly, spin-up remains up under E(2) in (0,1/2), and spin-down remains down under E(2) in (1/2,0).

Multiplying v_d and v_u by $R^{-1}(\theta,\varphi)B_C(\alpha)$, with C as usual A or B, we obtain the vectors

$$w_{A-1} = \exp(\alpha)\begin{pmatrix} \exp(-i\varphi)(1-\cos\theta)/2 \\ -\sin\theta/\sqrt{2} \\ \exp(i\varphi)(1+\cos\theta)/2 \end{pmatrix} \quad \textit{and} \tag{75a}$$

$$w_{B+1} = \exp(\alpha)\begin{pmatrix} \exp(-i\varphi)(1+\cos\theta)/2 \\ \sin\theta/\sqrt{2} \\ \exp(i\varphi)(1-\cos\theta)/2 \end{pmatrix} \tag{75b}$$

which satisfy the equations

$$\mathbf{J}\cdot\mathbf{p}\ w_{Cs} = spw_{Cs}\,,\quad s = \mp 1\ . \tag{76}$$

When p equals p^0, these equations are covariant under Lorentz transformations: helicity is then a particle property, quite like it is with spin-one-half particles of zero mass. The vectors with helicity are called w_{As} and w_{Bs} to distinguish them from the vectors produced by application of $R^{-1}(\theta,\varphi)B_cR(\theta,\varphi)$. The latter shall be called v_{As} and v_{Bs}. Recall the spinors χ_{As}, χ_{Bs} and ξ_{As}, ξ_{Bs} of eqs. (56) and (12).

The expression of Lorentz transformed vectors $w'_{Cs}(\alpha',\theta',\varphi')$ in terms of Wigner rotations and vectors $w_{Cs}(\alpha',\theta',\varphi')$ follows the same recipe as in the case of Lorentz transformed spinors χ'_{Cs}, see eqs. (52) and (53).

4.2 MASSIVE PARTICLES

The six vectors v_{As} and v_{Bs}, s assuming the values 1, 0, -1, can serve as spin functions of massive particles. Their components shall be pointed out by superscripts. Equations which the components satisfy may be derived by the same basic procedure which we employed to obtain eqs. (22) and ultimately the Dirac equation. The $p^{m,n}$ are familiar from eqs. (21). We start with the equation

$$-p^{1/2,1/2}\,v_B^{o} + \sqrt{2}\,p^{1/2,-1/2}\,v_B^{1} = k\,\Phi_{II}^{1/2,1/2} \tag{77}$$

with k a proportionality constant. The first superscript on $p^{m,n}$ and $\Phi_{II}^{m,n}$, referring to A-space, is a mere spectator of the summation. The second superscript on $p^{m,n}$ and the superscript on v_B^k combine to produce the superscript one half on $\Phi_{II}^{m.n}$ in B-space. To see if the addition of angular momenta one-half and one indeed produces an angular momentum one-half, we need to make sure that the coefficients on the left are correctly chosen. This is most simply done by writing the following vector equation, highlighting the transformation behavior of eq. (77),

$$\begin{pmatrix}1\\0\end{pmatrix}\begin{pmatrix}0\\1\\0\end{pmatrix} + \sqrt{2}\begin{pmatrix}0\\1\end{pmatrix}\begin{pmatrix}1\\0\\0\end{pmatrix} = \sqrt{3}\begin{pmatrix}1\\0\end{pmatrix} \tag{78a}$$

and operating on it with the ladder operator and with the diagonal operator

$$J_1 + iJ_2 = \begin{pmatrix}0 & 1\\0 & 0\end{pmatrix} + \sqrt{2}\begin{pmatrix}0 & 1 & 0\\0 & 0 & 1\\0 & 0 & 0\end{pmatrix} \tag{78b}$$

$$J_3 = (1/2)\begin{pmatrix}1 & 0\\0 & -1\end{pmatrix} + \begin{pmatrix}1 & 0 & 0\\0 & 0 & 0\\0 & 0 & -1\end{pmatrix} . \tag{78c}$$

Multiplication of eq. (77) by a matrix proceeds in the normal fashion. In the case of eq. (78a) one needs to multiply the column spinors and vectors by matrix columns. For consider some vector

$$A = a^k V_k \tag{79a}$$

the V_k being basis vectors, e.g., column vectors as in eq. (71). Transform

$$A' = a'^k V_k = a^k V'_k \tag{79b}$$

$$a'^k = t^k{}_l a^l \tag{79c}$$

and thus

$$V'_k = t'^{\,l}_{\,k} V_l \,. \tag{79d}$$

Guided by eq. (79d), when we apply the ladder operator of eq. (78b) to the products of vectors on the left of eq. (78a) we indeed get zero. J_3 then produces the eigenvalue one half.

- The general formula for the addition of angular momenta was derived by Wigner (Wigner 1931, Racah 1942). It may by now be obvious, that the addition of angular momenta follows from the reduction of the direct product of two representations of the three-dimensional rotation group. -

In eq. (77) the superscripts which are added to one another refer to B-space. The unvarying superscript refers to A-space. Thanks to Weyl's unitarian trick, the two together describe Lorentz covariance. (Previously, see eqs. (22), we had supplied the spinors corresponding to the vectors here with two superscripts, one of which was always zero. This was done to emphasize the representations with which we were dealing, (1/2,0) and (0,1/2). It should no longer be necessary to be that explicit.) Eq. (77) has a partner,

$$-p^{\,1/2,1/2}\, v_A^{\,o} + \sqrt{2}\, p^{\,-1/2,1/2}\, v_A^{\,1} = k\Phi_I^{1/2,1/2} \,. \tag{80}$$

The sum is now over the indices referring to A-space. The remaining equations are

$$\begin{aligned} -p^{\,+1/2,-1/2}\, v_A^{\,o} + \sqrt{2}\, p^{\,-1/2,-1/2}\, v_A^{\,1} &= k\,\Phi_I^{+1/2,-1/2} \\ p^{\,-1/2,+1/2}\, v_A^{\,o} - \sqrt{2}\, p^{\,1/2,1/2}\, v_A^{\,-1} &= k\,\Phi_I^{-1/2,+1/2} \\ p^{\,-1/2,-1/2}\, v_A^{\,o} - \sqrt{2}\, p^{\,+1/2,-1/2}\, v_A^{\,-1} &= k\,\Phi_I^{-1/2,-1/2} \end{aligned} \tag{81}$$

and the analogous equations for v_B , where the values of the superscripts are exchanged, and where $\mathbf{\Phi}$ gets the subscript II.

We already know the expressions $p^{m,n}$ in terms of Cartesian components p_k and p^0, see section 3.1.2. The components $\Phi_I^{m,n}$ and $\Phi_{II}^{m,n}$ are made up in the same way of Cartesian components Φ_{Ik} and Φ_{IIk} and components Φ_I^0 and Φ_{II}^0. First we rewrite the eight equations above as

$$(\mathbf{J}\cdot\mathbf{p}\pm p^{0}))v_{A,B} = k\Phi_{I,II} \tag{82a}$$

and

$$-p^{1}v_{A,B}^{-1} + p^{o}v_{A,B}^{o} - p^{-1}v_{A,B}^{1} = \pm k\Phi_{I,II}^{0} \tag{82b}$$

p, $v_{A,B}$ and $\Phi_{I,II}$ having spherical components, or with Cartesian components, in the coordinate representation,

$$\nabla\times\mathbf{v}_{\mathbf{A,B}} \pm i\dot{\mathbf{v}}_{\mathbf{A,B}} = k\boldsymbol{\Phi}_{I,II} \tag{83a}$$

and

$$\nabla\cdot\mathbf{v}_{\mathbf{A,B}} = \pm ik\Phi_{I,II}^{0} \ . \tag{83b}$$

To economize, the equations in v_A and Φ_I and in v_B and Φ_{II} have been combined. Note, only 3-vectors are bold and only when definitely in Cartesian coordinates. Upon judiciously differentiating, the curl equations and the divergence equations become the subsidiary conditions

$$\nabla\cdot\boldsymbol{\Phi}_{I,II} + \dot{\Phi}_{I,II}^{0} = 0 \ . \tag{84}$$

We continue in this vein by combining $p^{m,n}$ and $\Phi_I^{m,n}$ to yield v_A^{l} ,

$$p^{+1/2,+1/2}\Phi_I^{+1/2,-1/2} - p^{+1/2,-1/2}\Phi_I^{+1/2,+1/2} = k' v_A^1$$

$$\begin{aligned} &p^{-1/2,+1/2}\Phi_I^{+1/2,-1/2} + p^{+1/2,+1/2}\Phi_I^{-1/2,-1/2} \\ &-(p^{-1/2,-1/2}\Phi_I^{+1/2,+1/2} + p^{+1/2,-1/2}\Phi_I^{-1/2,+1/2}) = k' v_A^o \end{aligned} \tag{85}$$

$$p^{-1/2,+1/2}\Phi_I^{-1/2,-1/2} - p^{-1/2,-1/2}\Phi_I^{-1/2,+1/2} = k' v_A^{-1} \,.$$

and similarly for Φ_{II} and v_B. Antisymmetrization in B or A, giving zero-spin, in the presence of constant superscripts 1/2 in A or B, respectively, results in $v^1_{A,B}$. Likewise for $v^{-1}_{A,B}$, except that the constant superscripts are now both -1/2. To obtain $v^o_{A,B}$ we antisymmetrize the superscripts referring to one spin-space, A or B and symmetrize the others. The symmetrization, recall from quantum mechanics, gives spin-one with J_3 equal to zero. The six equations can be collected as

$$(-\mathbf{J}\cdot\mathbf{p} \pm p^0)\Phi_{I,II} \mp P\Phi^0_{I,II} = k' v_{A,B} \,. \tag{86}$$

P is the momentum vector with spherical components. Again, Cartesian components and the coordinate representation make for more familiar looking equations

$$-\nabla\times\mathbf{\Phi}_{I,II} \pm i(\nabla\Phi^0_{I,II} + \dot{\mathbf{\Phi}}_{I,II}) = k' \mathbf{v}_{A,B} \,. \tag{87}$$

4.2.1 Second Rank Tensors

Before taking the final steps in the derivation of equations in $\mathbf{v}_{A,B}$ and in $\mathbf{\Phi}_{I,II}$, $\Phi^0_{I,II}$, respectively, it is useful to establish a connection between the 3-vectors $v_{A,B}$ and antisymmetric tensors. The (complex) Cartesian components $v^k_{A,B}$ transform under three-dimensional rotations like the A^k and B^k, respectively, and consequently like the J^k and like the iK^k and $-iK^k$, respectively. (In the case (0,1) $-iK^k$ is the same as iK^k in the case (1,0). Both are equal to J^k). The $v^k_{A,B}$ may also be subjected to actual three-by-three Lorentz transformations generated by the K^k of (1,0) and (0,1), respectively; so may the J^k and $\pm$ iK^k ! But we know that the vectors **J** and **K** are the axial and polar vector constituents of the angular-momentum tensor $J^{\mu\nu}$. Therefore we can define the second-rank tensors

$$F_C^{lm} = v_C^k \text{ and } F_C^{0k} = \mp i v_C^k , \; C = A \text{ and } B, \text{ respectively} . \qquad (88a)$$

Thus, the behavior of the 3-vectors $v_{A,B}$ under, say, a standard Lorentz transformation along the third axis may be described by either of the matrices

$$\begin{pmatrix} \cosh\alpha & \mp i\sinh\alpha & 0 \\ \pm i\sinh\alpha & \cosh\alpha & 0 \\ 0 & 0 & 1 \end{pmatrix} \; or \; \begin{pmatrix} 1 & 0 & 0 & 0 \\ 0 & 1 & 0 & 0 \\ 0 & 0 & \cosh\alpha & \sinh\alpha \\ 0 & 0 & \sinh\alpha & \cosh\alpha \end{pmatrix} \qquad (88b)$$

the first multiplying v_C^k, products of the second multiplying $F_{C\mu\nu}$ in the usual manner

$$F'^{\mu\nu}_C = l^{\mu w} l^{\nu x} F_{Cwx} \; .$$

When they act on Cartesian components, as they do above, the three-by-three representations are obvjously complex orthogonal, and thus

$$v_C^k = v_{Ck} \; . \qquad (88c)$$

Furthermore

$$v_A^k \sim v_B^{k*} \; . \qquad (88d)$$

Eqs. (88a) are an expression of the isomorphism of the Lorentz and the three-dimensional complex orthogonal groups.

We now introduce the real tensor, without loss of generality

$$F^{lm} = (v_A^k + v_B^k)/2\,, F^{0k} = i(v_B^k - v_A^k)/2 \tag{89a}$$

or

$$F^{\mu\nu} = (F_A^{\mu\nu} + F_B^{\mu\nu})/2\ . \tag{89b}$$

We also introduce the "dual tensor" - there are two representations to take care of

$$F^{D\mu\nu} = \varepsilon^{\mu\nu wx} F_{wx}/2\,, \varepsilon^{1230} = 1\ . \tag{90}$$

$\varepsilon^{\mu\nu wx}$ is the "Levi-Civita tensor". It stays plus one under even permutations of the indices, and becomes minus one under odd permutations. Thus

$$F^{Dlm} = F^{0k} \text{ and } F^{D0k} = -F^{lm}\ . \tag{91}$$

The dual tensor is of particular interest in the massless case.

Going into the coordinate representation, eqs. (83) can now be cast into the manifestly covariant forms

$$\frac{\partial F_A^{\mu\nu}}{\partial x^\nu} = k\Phi_I^\mu\ ,\ \frac{\partial F_B^{\mu\nu}}{\partial x^\nu} = k\Phi_{II}^\mu\ . \tag{92a}$$

If we split the "potential functions" into a vector and an axial or pseudo vector, we obtain

$$\frac{\partial F^{\mu\nu}}{\partial x^\nu} = k\,\Phi_V^\mu\ ,\ \frac{\partial F^{D\mu\nu}}{\partial x^\nu} = k\,\Phi_P^\mu\ . \tag{92b}$$

Similarly, eqs. (87) can be cast into

$$-k'F_{\mu\nu} = \frac{\partial \Phi_{V\nu}}{\partial x^{\mu}} - \frac{\partial \Phi_{V\mu}}{\partial x^{\nu}} - \varepsilon_{\mu\nu w x} \frac{\partial \Phi_{Px}}{\partial x^{w}} \tag{93a}$$

$$-k'F^{D}_{\mu\nu} = \varepsilon_{\mu\nu\kappa\kappa} \frac{\partial \Phi_{Vk}}{\partial x^{\omega}} + \frac{\partial \Phi_{P\nu}}{\partial x^{\mu}} - \frac{\partial \Phi_{P\mu}}{\partial x^{\nu}} \ . \tag{93b}$$

4.2.2 Equations for Massive Particles and Fields

Because of the Pauli principle, we are generally not concerned with classical spin-one-half fields. But in the case of spin-one we are concerned with classical fields. As a matter of fact, the equations to follow are usually looked upon as field equations. However, from the general solutions one can extract parts which may serve as wave functions, and which shall be of particular use to us in transition amplitudes. See, e.g., the end of section 3.4.1.

In an analogy to the Dirac Equation, from our point of view the equation for the 4-spinor made up of the 2-spinors ξ_A and ξ_B , we construct from eqs. (82) and (86) the equation for the "6-vector" made up of the 3-vectors v_A and v_B. To describe one kind of particle or field, we equate the contributions from the pseudovector potential, $\nabla\Phi^0_P + \dot{\mathbf{\Phi}}_P$ *and* $\nabla\times\mathbf{\Phi}_P$, to zero:

$$\begin{aligned}(-\mathbf{J}\cdot\mathbf{p}+p^{0})(\mathbf{J}\cdot\mathbf{p}-p^{0})v_B &= -P\mathbf{p}\cdot\mathbf{v}_B + kk'v_A \\ (-\mathbf{J}\cdot\mathbf{p}-p^{0})(\mathbf{J}\cdot\mathbf{p}+p^{0})v_A &= -P\mathbf{p}\cdot\mathbf{v}_A + kk'v_B \\ kk' &= m^2 \ .\end{aligned}$$

With $\mathbf{J}\cdot\mathbf{p}\,\mathbf{J}\cdot\mathbf{p} = p^2 - P\mathbf{p}\cdot$ the quantum mechanical version of the curl-curl decomposition, we obtain the equation

$$(-2\mathbf{J}\cdot\mathbf{p}(\mathbf{J}\cdot\mathbf{p}\mp p^{0}) + p^2 - (p^{0})^2)\begin{pmatrix} v_B \\ v_A \end{pmatrix} = m^2\begin{pmatrix} v_A \\ v_B \end{pmatrix} \ . \tag{94}$$

The 3-vectors which are not bold, v_A, v_B and P, are to be viewed as column vectors, with spherical or Cartesian components, depending on the representation one uses for the matrices J_k. The equation is made up of a pair of three second-order equations, apparently coupled by the particle's mass. If the mass term vanishes, we get, for $p^2 = (p^0)^2$, first-order equations. In eq. (94) $(p^0)^2 - p^2$ is, for the sake of generality, not equated to m^2 (Weinberg 1964). Like the Dirac equation, eq. (94) is symmetric about (0,0). But because it connects subspaces belonging to representations two units of angular momentum apart, (1,0) and (0,1), it is of second order.

From eqs. (83) and (87), and $\Phi_p^\mu = 0$, we can obtain

$$\nabla^2 F^{\mu\nu} - \ddot{F}^{\mu\nu} = m^2 F^{\mu\nu} \tag{95}$$

plus the subsidiary or initial condition

$$\nabla \cdot \boldsymbol{B} = 0 \;,\; B^k = F^{lm} \;. \tag{96}$$

With $m = 0$, these equations are familiar from electromagnetic theory. The only restriction we come up with is the vanishing of the divergence of F^{lm}. Keeping the pseudo-vector potential instead of the vector potential yields a zero-divergence of the polar 3-vector part of the tensor.

Of greater use, as far as we know, is the set of equations for Φ^μ

$$\frac{\partial}{\partial x^\nu}\left(\frac{\partial \Phi^\nu}{\partial x_\mu} - \frac{\partial \Phi^\mu}{\partial x_\nu}\right) = -m^2 \Phi^\mu \tag{97}$$

the "Proca Equation" (Proca 1936). It follows from eqs. (92b) and (93a) and, as usual, from setting Φ_p^μ equal to zero. Because of the mass term, the condition

$$\frac{\partial \Phi^\mu}{\partial x^\mu} = 0 \tag{98}$$

follows; see eqs. (84). Consequently, the potential also satisfies Klein-Gordon equations, plus a subsidiary condition. In the rest frame Φ^μ has spin one.

The superiority of the Proca equation over eqs. (96) arises once interactions are incorporated.

4.2.3 The Vector Potential

As with 2-spinors and 3-vectors, we standardize 4-vectors. In the helicity representation this gives

$$\Phi_s(p) = \mathfrak{R}^{-1}(\theta,\varphi)\mathfrak{Z}(\alpha)u_s \tag{99}$$

with

$$\mathfrak{R}(\theta,\varphi) = \wp'^{-1}\begin{pmatrix} & & & 0\\ & R(\theta,\varphi) & & 0\\ & & & 0\\ 0 & 0 & 0 & 1\end{pmatrix}\wp' \tag{100a}$$

$$R(\theta,\varphi) = \begin{pmatrix}\cos\theta & 0 & -\sin\theta\\ 0 & 1 & 0\\ \sin\theta & 0 & \cos\theta\end{pmatrix}\begin{pmatrix}\cos\varphi & \sin\varphi & 0\\ -\sin\varphi & \cos\varphi & 0\\ 0 & 0 & 1\end{pmatrix} \tag{100b}$$

$$\mathfrak{Z}(\alpha) = \wp'^{-1}\begin{pmatrix}1 & 0 & 0 & 0\\ 0 & 1 & 0 & 0\\ 0 & 0 & \cosh\alpha & \sinh\alpha\\ 0 & 0 & \sinh\alpha & \cosh\alpha\end{pmatrix}\wp' \tag{101}$$

and

$$\wp' = \begin{pmatrix} & & & 0\\ & U' & & 0\\ & & & 0\\ 0 & 0 & 0 & 1\end{pmatrix}, \; U' \textit{ given by eq.}(69)\;. \tag{102}$$

The u_s are four component basis vectors, s running from u (up) to o (J_3 equal to zero), d (down) and t (timelike, J and J_3 equal to zero); compare to eqs. (71).

We transform to the coordinate representation, the coefficients $a^s(p)$ multiplying the standard vectors. Here we use superscripts s to denote spherical components. Cartesian components, see eqs. (104), have subscripts. Since it is common practice to represent the potential by its Cartesian components, we do so, transforming from the spherical components in which the standard vectors appear:

$$\Phi^\mu(x^\nu) = (2\pi)^{-3/2} \int (d^3p/\sqrt{(2p^o)}) \begin{pmatrix} & & & 0 \\ & R-1(\theta,\varphi) & & 0 \\ & & & 0 \\ 0 & 0 & 0 & 1 \end{pmatrix}^\mu B(\alpha)$$

$$\wp \left[\begin{pmatrix} a^{+1}(p) \\ a^o(p) \\ a^{-1}(p) \end{pmatrix} \exp(ip^\nu x_\nu) + \begin{pmatrix} a^{-1}(p) \\ a^o(p) \\ a^{+1}(p) \end{pmatrix}^* \exp(-ip^\nu x_\nu) \right] . \tag{103}$$

B(α) is the boost operating on Cartesian components, see eq. (101). $\wp$ differs from $\wp'$ in that its submatrix U differs from U′ in the sign of the first column. Thus the three a^s in eq. (103) are

$$a^{+1} = (a_1 - ia_2)/\sqrt{2}\,,\; a^o = a_3\,,\; a^{-1} = (a_1 + ia_2)/\sqrt{2}\;. \tag{104}$$

The first eigenvector in eq. (69), which, instead of as a column, may be written $-(e_1 + ie_2)/\sqrt{2}$, is associated with the spherical component $-(a_1 - ia_2)/\sqrt{2}$. The other two eigenvectors, and the components, or coefficients, associated with them, have plus signs in front. In this way the matrix elements of the usual ladder operators of quantum mechanics are positive. This is in agreement with Wigner (Wigner 1931). Consequently, whenever we face equations directly or indirectly involving the standard angular-momentum matrices of eqs. (70), we use the coefficients just mentioned. However, if we deal with Cartesian components of fields and potentials, the spherical coefficients appearing in them will, as is customary, be those of eqs. (104).

- The quadratic form $a_k a_k$ is an invariant of the three-dimensional complex orthogonal group. We can express it as $U_{kr} a^r U_{ks} a^s$, r and s going from -1 to o to +1. Thus we obtain the relevant metric tensor

$$T = U^T U = \begin{pmatrix} 0 & 0 & 1 \\ 0 & 1 & 0 \\ 1 & 0 & 0 \end{pmatrix} . \tag{105}$$

In eq. (103) a scalar $a_o^o(p)$ does not appear: one of the results of connecting the subspace of (1,0) and (0,1), respectively, to that of (1/2,1/2) by operating on the former with $p^{m,n}$, is eq. (82b).

$$For \qquad k \neq 0:\ \Phi^0_{I,II} = 0 \ if\ p^s = 0\,;\, s = 1, o, -1\ .$$

(1,0) and (0,1) contain no scalar in 3-space, and thus the only such scalars which can be formed from $p^{m,n}$, which also belongs to (1/2,1/2), and $v^k_{A,B}$ are the products $\mathbf{p} \cdot \mathbf{v}_{A,B}$. - The use of the same coefficients in the positive and negative energy terms implies a self-charge-conjugate, or neutral, spin-one-particle. -

The $\sqrt{(2p^o)}$ in the denominator makes for simple expressions for observables like the radiation-field-energy and momentum in terms of the coefficients $a^s(p)$ and p^μ [A.8] (Bogoliubov 1959).

A comment: the matrix $\wp$ in eq. (103) may, of course, be relieved of its fourth column.

The massive fields are easily obtained with the help of eqs. (93a).

4.3 MASSLESS FIELDS AND POTENTIALS

Using the standard vectors of eqs. (75) and, as with the vector potential, converting to Cartesian components, the two 3-vectors describing a massless spin-one field may be written

$$w^{k}_{\pm 1}(x^{\mu}) = \pm(2\pi)^{-3/2} \int (d^3p/\sqrt{(2p^0)})R^{-1k}(\theta,\varphi)B_{B,A}(\alpha)$$

$$1/\sqrt{2}\begin{pmatrix} 1 & 0 & 1 \\ i & 0 & -i \\ 0 & 0 & 0 \end{pmatrix}\left[\begin{pmatrix} a^{+1} \\ 0 \\ a^{-1} \end{pmatrix} \exp(ip^{\mu}x_{\mu}) - \begin{pmatrix} a^{-1} \\ 0 \\ a^{+1} \end{pmatrix}^{*} \exp(-ip^{\mu}x_{\mu})\right] \tag{106}$$

B_B, a^{+1} and a^{-1*} go with w_{+1}, B_A, a^{-1} and a^{+1*} go with w_{-1}. The boosts are given by the first of expressions (88b). Eqs. (106) satisfy

$$w^{k*}_{+1}(x^{\mu}) = w^{k}_{-1}(x^{\mu}) \ . \tag{107}$$

The field components can be obtained from

$$w^{k}_{\pm 1}(x^{\mu}) = \partial A^{m}_{\pm 1}/\partial x^{l} - \partial A^{l}_{\pm 1}/\partial x^{m} \tag{108a}$$

the potentials $A^{\nu}_{\pm 1}(x^{\mu})$ resulting from the massive potential of eq. (103) by setting the two lower $a^{s}(\mathbf{p})$ and the two upper $a^{s}(\mathbf{p})$, respectively, equal to zero. To see this, we combine the rotation matrix and eigenvectors into "polarization vectors"

$$\wp_{\pm 1}(\mathbf{p}) = (1/\sqrt{2})\, R^{-1}(\theta,\varphi)\begin{pmatrix} 1 \\ \pm i \\ 0 \end{pmatrix} . \tag{109a}$$

It is convenient to rewrite these equations as

$$\wp^{k}_{\pm 1}(\mathbf{p}) \equiv e^{k}_{\pm 1}(\mathbf{p}) = R^{-1k}_{\ \ l}(\theta,\varphi)\, e^{l}_{\pm 1}$$

in terms of the spherical unit vectors

$$e_{\pm1} = (\mathbf{e}_{\pm1} \pm i\mathbf{e}_2)/\sqrt{2}$$

because we have

$$p\mathbf{e}_{\pm1} = \pm i\,\mathbf{p} \times \mathbf{e}_{\pm1} \ , \ \ \mathbf{p} = p\mathbf{e}_3$$

and thus

$$p\,\mathbf{e}_{\pm1}(\mathbf{p}) = \pm i\,\mathbf{p} \times \mathbf{e}_{\pm1}(\mathbf{p}) \tag{109b}$$

for any momentum. Consequently, the integrands in eqs. (106) indeed assume the form

$$\begin{aligned} i(2p^0)^{-1/2}(p^l \wp^m_{\pm1}(\mathbf{p}) - p^m \wp^l_{\pm1}(\mathbf{p}))\,[a^{\pm1}(\mathbf{p}) \exp(ip^\mu x_\mu) \\ - a^{+1\,*}(\mathbf{p}) \exp(-ip^\mu x_\mu)] \end{aligned} \tag{110}$$

which is expected from eqs. (108a). The momentum magnitude, p, comes from the boosts, which, if taken through the matrix of column eigenvectors, results in the factor

$$\exp(\alpha) = p/p_i$$

multiplying $a^{+1}(\mathbf{p})$ and $a^{-1}(\mathbf{p})$, and their complex conjugates; the initial momentum was set equal to one.

The 6-vector of eqs. (106) and (110) may be expressed as a tensor via eqs. (88a), which, as is apparent from the discussion leading up to them, hold independently of any mass considerations.

Having shown that eqs. (108a) indeed hold, a comparison with eqs. (87) is indicated. There the 3-vectors $\mathbf{v}_A$ and $\mathbf{v}_B$ are given in terms of derivatives of the 4-vectors Φ_I and Φ_{II} , respectively. Again setting Φ_P equal to zero, eqs. (87) read

$$k'v^k_{A,B} = -(\partial\Phi^m/\partial x^l - \partial\Phi^l/\partial x^m) \pm i(\partial\Phi^0/\partial x^k - \partial\Phi^k/\partial x_0)$$
$$\text{or} \qquad k'F^{\mu\nu}_{A,B} = -(\partial\Phi^\nu/\partial x_\mu - \partial\Phi^\mu/\partial x_\nu) \mp i\,\varepsilon^{\mu\nu\omega\chi}\,\partial\Phi_\chi/\partial x^\omega \ . \qquad (87')$$

For zero mass each 3-vector has its own potential. Remembering that we always changed symbols when going over to the massless case, this means w_{-1} has A_{-1} and w_{+1} has A_{+1}. How the $A_{\pm 1}$ are related to Φ is mentioned below eqs. (108a). Eqs. (108) correspond to the above equations, except they do not contain the second terms on the right.

Let us now look at a particular momentum, say $p = p_3$. The four by four boost along the third axis and the matrix $\wp$ of eigenvectors combine to give zero third and fourth components of $A_{\pm 1}$. The radiation potential reduces to

$$A^k_{\pm 1}(x^\mu) = (2p^0)^{-1/2}\, U^k_{\pm 1}(a^{\pm 1}(\mathbf{p}) \exp(ip^\mu x_\mu)$$
$$+ (a^{\mp 1 *}(\mathbf{p}) \exp(-ip^\mu x_\mu)) \ .$$

We derive

$$k'\, w^2_{\pm 1} = -\,\partial A^1_{\pm 1}/\partial x^3 \mp i\,\partial A^2_{\pm 1}/\partial x^0$$
$$k'\, w^1_{\pm 1} = \partial A^2_{\pm 1}/\partial x^3 \mp i\,\partial A^1_{\pm 1}/\partial x^0 \qquad (108b)$$

and verify that in each equation the first and second terms on the right including signs are equal! Since the $w^k_{\pm 1}$ satisfy homogeneous equations, it is up to us with what constant to connect fields and potentials. The choice $k' = -1$ will lead us to the standard relation between fields and potentials, since the $w_{\pm 1}$ are then twice what they are according to eqs. (108a).* After all, the $F^{\mu\nu}$ are given by $F^{\mu\nu}_A + F^{\mu\nu}_B$ times one half, and thus

$$F^{\mu\nu} = \partial(A^\nu_{+1} + A^\nu_{-1})/\partial x^\mu - \partial(A^\mu_{+1} + A^\mu_{-1})/\partial x^\nu \ . \qquad (108c)$$

* With Weinberg's use of equations like eqs. (108a), one is lead to the $F^{\mu\nu}$ of eq. (108c) divided by two (Weinberg 1965, pp. 455-457).

F_A and F_B have doubled along with the $w_{\pm 1}$; see eqs. (88a).

In the massive case the terms involving the Levi-Civita tensor cancel, when F is formed from F_A and F_B. But apart from those terms, F_A and F_B are equal. (Remember, $\Phi_p = 0$.) Thus

$$F^{\mu\nu} = \partial\Phi^{\nu}/\partial x^{\mu} - \partial\Phi^{\mu}/\partial x^{\nu} .$$

Someone may object: recall the origin of eqs. (87), which is to be found in eqs. (85), which in turn are based on the reductions

$$(1/2,1/2) \otimes (1/2,1/2) \rightarrow (1,0) \; and \; (0,1) .$$

But in the massless case we do not have a subspace of the second representation on the left. So, eqs. (87) do not apply. To this one can answer: they do apply. The radiation potentials $A_{\pm 1}$ acquire longitudinal and scalar components when subjected to transformations in L_h, but the new terms leave the fields alone. For more of this see section 5.2.

4.3.1 Helicity, Polarization and Maxwell's Equations

The 3-vector-function

$$f_{+1}(x^{\mu}) = \begin{pmatrix} 1 \\ 0 \\ 0 \end{pmatrix} \exp(i(px_3 - p^0 x^0)) \tag{111a}$$

is a solution of the lower of eqs. (76). It describes positive helicity, since p is the momentum magnitude. In terms of the Cartesian unit vectors we have

$$\begin{aligned} f_{+1}(x^{\mu}) &= (\mathbf{e}_1 + i\,\mathbf{e}_2)/\sqrt{2}\, \exp(i(px_3 - p^0 x^0)) \\ &= (\mathbf{e}_1 \cos\varphi - \mathbf{e}_2 \sin\varphi)/\sqrt{2} + i(\mathbf{e}_1 \sin\varphi + \mathbf{e}_2 \cos\varphi)/\sqrt{2} \end{aligned} \tag{111b}$$

φ short for the x_3 and x^0 dependence. Thus positive helicity corresponds to left-circular polarization (Young 1968). Let us provide the column vector above with the amplitude, or coefficient, a^{+1}, see eq. (104), made up of

$$a_1 = (a_{1r} + i a_{1i})/\sqrt{2} \ \ and \ \ a_2 = (a_{2r} + i a_{2i})/\sqrt{2} \ .$$

Since the down-component vanishes, the amplitude a^{-1} is zero, i.e.,

$$a_{1r} = a_{2i}, a_{1i} = -a_{2r} \ . \tag{111c}$$

We rewrite the resulting vector in Cartesian components and multiply by p/√(2p⁰), to obtain, see eqs. (106) and (110),

$$w_{+1} = p\,U a^{+1}(\mathrm{p}) f_{+1}(x^\mu)/\sqrt{(2p^0)} = \frac{1}{2}\begin{pmatrix} B_1(x^\mu) - iE_1(x^\mu) \\ B_2(x^\mu) - iE_2(x^\mu) \\ 0 \end{pmatrix} \ . \tag{111d}$$

where, on account of eqs. (111c), and the choice of relative signs on the right side of the above equation, electric and magnetic fields are related by

$$B_1 = -E_2, E_1 = B_2 \ .$$

Inserting this solution in the lower of eqs. (76) (in the coordinate representation), with J_3 given by eq. (67), we obtain Maxwell's homogeneous equations for zero B_3 and E_3. Rotation of the coordinates, or the fields, gives all components of Maxwell's equations. Employing negative instead of positive helicity and the upper of eqs. (76) leads to the same result.
- Choosing in eq. (111d) a relative plus sign between E_k and B_k would simply be contrary to convention. - Note that according to eqs. (89a) $F^{0k} = E_k$ and $F^{lm} = B_k$. See also remarks preceding eq. (108c) and following eqs. (87′).

While weakly interacting currents of spinors distinguish sharply between helicities, see eq. (37), the coupling of the current to the electromagnetic potential does not. In weak interactions we would have a somewhat analogous situation, if we had massless Majorana neutrinos and a coupling without the $1\pm\gamma_5$.

The electromagnetic potential appearing in transition amplitudes (see the statement following eq. (35d)) consists of a polarization vector, see eq. (109a), the factor $1/\sqrt{(2p^0)}$, and either one of the functions $\exp(\pm ip^{\mu}x_{\mu})$. The product of the three is the factor of the coefficient $a^s(\mathbf{p})$ and $a^{s*}(\mathbf{p})$, respectively. - In the quantum field theory, these coefficients are the particle "destruction and creation operators." - In the classic theory we sum exponentials. Say we are interested in the potential description of a left-circularly-polarized traveling wave. We then sum the positive-frequency-piece of the integrand of $A_{+1}(x^{\mu})$ and the negative one of $A_{-1}(x^{\mu})$. The resulting potential, and thus also the field strength, is real. -

In general, the starred coefficient need not be the complex conjugate of its unstarred partner with the same s and **p**. If it is not, it is associated with the field of antiparticles. If it is, the field is self-charge-conjugate.

4.4 INHOMOGENEOUS TRANSFORMATIONS

We have been emphasizing the representations of the homogeneous Lorentz group and their subspaces. When we introduced the coordinate representation (see eq. (30a)), the reader may have wondered, if not more than manipulations of the group L_h was involved. There was. Making the constants p^{μ}, originating from boost and rotation (see eqs. (11c) and (12a)), the eigenvalues of the operators $-i\,\partial/\partial x_{\mu}$, extends the group to L_i, the "inhomogeneous Lorentz group" or "Poincare group". The operators are the generators of translations.

Strictly speaking, substituting specifically p^{μ} and m for angles and boost was merely a way of introducing a familiar 4-vector into the spinor components. Any identification with velocity, requiring space coordinates and a time coordinate with a measure, was unnecessary. As the reader should know, mathematically a standard Lorentz transformation or "antiboost" may be viewed as inclining two originally perpendicular axes towards one another. With axes x_3 and x^0 and their primed counterparts, the angle β between the x_3 and x_3' axes is given by

$$\tan\beta = \tanh\alpha$$

since the equation of the x_3' axis is

$$x^0 = x_3 \tanh\alpha \; .$$

Understanding is just as important as predicting.

Heisenberg

PART TWO

So far the main concern has been with global transformations in space-time, the wave and field functions they generate, and the equations these satisfy. The fundamental role for physics of the representations (A,0) and (0,B) of L_h for A and B equal to one-half and one became apparent.

In the following, local transformations are introduced in form of the gauge transformations, invariance under which is the well-known rational for the electromagnetic interaction. They are extended to include three-dimensional (local) gauge transformations in an "isospace", to provide a reason for both, the electromagnetic and weak interactions, and their unification. Unlike the by all accounts massless case of electromagnetism, the weak interactions require a scheme to generate mass. The well known scheme which accomplishes this is crucially dependent on the local character of the gauge transformations. While leaving Planck's electromagnetic quantum massless, it is remarkably successful in bestowing mass on the "new quanta". Spin-one states with helicity are miraculously converted into ordinary spin-one states. On the other hand, each charged-lepton mass is generated independently, in a manner of speaking "put in by hand." - This "electroweak theory" involves zero-mass neutrinos.

Some time is spent on taking a bird's eye view of the developments beginning with Heisenberg's isotopic spin, to go on to concepts like charge independence, strangeness, SU(3), generations, weak isospin and, presumably a major breakthrough, the color force.

At the end of Part Two, the concepts and formalism used in work on neutrino properties is discussed. There is the problem of the neutrino mass, and the question, whether neutrinos are of Dirac or Majorana type, or maybe of both types. In the areas of neutrino oscillations and double-beta decay, transformations between particle states are essential. The basics of some experiments are discussed, particularly as they affect calculations. The discussion of neutrino oscillations and of the particle-physics aspects of double-beta decay does not require previous acquaintance with these phenomena.

Chapter 5

GAUGE TRANSFORMATIONS

We shall now look at "local gauge transformations" or "gauge transformations of the second kind" (Pauli 1940), and some of their consequences. They do not involve space-time dependent coordinate transformations - otherwise we might be lead to general relativity, not our intent - but rather space-time dependent phase transformations:

$$\psi(x^{\mu}) \rightarrow \psi'(x^{\mu}) = \exp(i\varphi(x^{\mu}))\,\psi(x^{\mu}) \ . \tag{112a}$$

The demand of invariance under such a transformation is properly based on the fundamental argument that ψ' is just as permissible as ψ !

Consider the Dirac equation. To render the equation invariant under the transformation (112a), one needs to have

$$D\psi(x^{\mu}) \rightarrow \exp(i\varphi(x^{\mu}))\,D\psi(x^{\mu}) \ . \tag{112b}$$

$$D = \gamma^{\mu}\partial/\partial x^{\mu} + m + X \tag{113a}$$

where X is a quantity, which transforms along with ψ in a way guaranteeing the condition (112b), i.e.,

$$X = -iq\gamma^{\mu}V_{\mu} \rightarrow -iq\gamma^{\mu}(V_{\mu} + (1/q)\partial\varphi/\partial x^{\mu}) \ . \tag{113b}$$

We are familiar with the emerging transformation of a "4-potential" from electromagnetic theory. There the 4-potential, A_{μ} , satisfies the equation

$$\frac{\partial}{\partial x^{\nu}}(\partial A^{\nu}/\partial x_{\mu} - \partial A^{\mu}/\partial x_{\nu}) = j^{\mu} \tag{114}$$

and the field $F^{\mu\nu} = \partial A^{\nu}/\partial x_{\mu} - \partial A^{\mu}/\partial x_{\nu}$ is obviously invariant under such a transformation. The homogeneous equation holds for the radiation potential. It is the Proca equation, eq. (97), for zero mass. See also section 5.3. Clearly, there is no a priori reason why we cannot replace the zero on the right by a 4-vector, which has zero divergence.

5.1 THE PROCA EQUATION WITH SOURCES IN THE LIMIT OF SMALL MASS

Planck's radiation law, in beautiful agreement with observation, contains a factor two due to the two polarization states of electromagnetic radiation, or two spin states of the photon. One still hears the statement made that this proves that the photon has no mass. - We had concluded that the representations (1,0) and (0,1) describe either three spin states or two helicity states.

Would the three spin states of a massive vector-particle really preclude it from being responsible for black-body-radiation, no matter how small its mass? In analogy to eq. (114), include a source term in the Proca equation describing a massive vector field,

$$\frac{\partial}{\partial x^{\mu}}(\partial \Phi^{\nu}/\partial x_{\mu} - \partial \Phi^{\mu}/\partial x_{\nu}) + m^{2}\Phi^{\mu} = j^{\mu}$$

frequently written

$$-\Box\Phi^{\mu} + \frac{\partial}{\partial x_{\mu}}\frac{\partial \Phi^{\nu}}{\partial x^{\nu}} + m^{2}\Phi^{\mu} = j^{\mu}\ . \tag{115}$$

This equation is not gauge invariant like eq. (114)!

There are no terms $\partial^2\varphi^{\mu}/(\partial x^{\mu})^2$ (no sum). - Einstein's Equations do not contain the terms $\partial^2 g^{\mu\nu}/(\partial x^{\mu})^2$ (no sum) (Adler 1975)! - Consequently, the pertinent Cauchy initial-value-problem involves only three of the four eqs. (115),

$$\ddot{\mathbf{\Phi}} + \nabla\times\nabla\times\mathbf{\Phi} + \nabla\dot{\Phi}^0 + m^2\mathbf{\Phi} = \mathbf{j}\ . \tag{116}$$

with $\mathbf{\Phi}$ and $\mathbf{j}$ the 3-vector parts of Φ^μ and j^μ respectively. - Note that in eq. (103), which gives the solution of the homogeneous Proca equation, only three Fourier coefficients appear for every $\mathbf{p}$. - The fourth equation,

$$\nabla\cdot(\nabla\Phi^0 + \dot{\mathbf{\Phi}}) - m^2\Phi^0 = -j^0 \tag{117}$$

becomes a condition on the initial conditions, since its time derivative equals the divergence of the vector equation. The latter equality contains the equation

$$m^2(\nabla\cdot\mathbf{\Phi} + \partial\Phi^0/\partial x^0) = \nabla\cdot\mathbf{j} + \partial j^0/\partial x^0\ . \tag{118}$$

Thus with nonvanishing mass the current is in general not conserved. Let us see, what effect this has on the vector equation, eq. (116). We express the potential as a Fourier integral,

$$\Phi^\mu(x^\nu) = (2\pi)^{-3/2}\int \exp(i\mathbf{p}\cdot\mathbf{r})\Phi^\mu(p,x^0)d^3p \tag{119}$$

and do the same for the current density. Using these integrals, and substituting in eq. (116) the Φ^0 from eq. (117), yields

$$(p^2 + \partial^2/(\partial x^0)^2)\Phi_k + ip_k[i\mathbf{p}\cdot\mathbf{\Phi} + (\partial j^0/\partial x^0 + i\mathbf{p}\cdot\partial^2\mathbf{\Phi}/(\partial x^0)^2)/(p^2+m^2)] + m^2\Phi_k = j_k\ . \tag{120a}$$

Introducing the components

$$\Phi_k' = (\delta_{kl} - p_kp_l/(p^2 + m^2))\Phi_l \tag{121}$$

and likewise with j_k , simplifies this equation,

$$(p^2 + \partial^2/(\partial x^0)^2)\,\Phi_k' + m^2\Phi_k' = j_k' - (i/(p^2+m^2))p_k(ip_l j_l + \partial j^0/\partial x^0)\,. \qquad (120b)$$

Approximating

$$p_k p_l/(p^2+m^2) \simeq p_k p_l(1/p^2 - m^2/p^4) \qquad (122)$$

provides us with separate equations for the transverse and longitudinal potentials (Feldman (1977),

$$\Phi^k_{tr} = (\delta_{kl} - (p^k p_l/p^2))\Phi^l\,,\ \Phi^k_{lon} = (p^k p_l/p^2)\Phi^l\,. \qquad (123)$$

Φ^k_{lon} is obviously proportional to p^k, and the vector $\mathbf{\Phi}_{tr}$ is consequently perpendicular to the momentum. In other words, Φ^k_{lon} has zero helicity, while Φ^k_{tr} has in general a combination of helicities plus one and minus one. In eq. (120b) we perform the approximation on Φ_k' , do the same for j_k' , and obtain

$$(p^2 + \partial^2/(\partial x^0)^2)\Phi^k_{tr} = j_{tr}\,,\ (p^2 + \partial^2/(\partial x^0)^2)(m^2/p^2)\Phi^k_{lon} = (m^2/p^2)j^k_{lon}$$
$$+ (p^k/p^2)(1 - m^2/p^2)(p_l j^l - i(\partial j^0/\partial x^0))\,. \qquad (124a,b)$$

The meaning of j_{tr} and j_{ion} can be inferred from eqs. (123). In the second equation there is a common factor mp^k/p^2, which we eliminate. Fourier transforming back to 3-space, we end up with the equations

$$\Box\Phi^k_{tr} = -j^k_{tr} \ \textit{and}\ \Box\Phi_s = -(\partial j^\mu/\partial x^\mu)/m + (\partial j^0_s/\partial x^0)/m\,. \qquad (125a,b)$$

The scalar potential

$$\Phi_s(x^\mu) = (2\pi)^{-3/2} i\, m \int \exp(i\mathbf{p}\cdot\mathbf{r})p_k\,\Phi^k(p,x^0)/p^2)d^3p$$

would only then not noticeably contribute to the emission or absorption of black body radiation, if the right side of eq. (125b) were to go to zero with m.* The charge density j_s^0 is the Fourier transform of

$$m^2 j^0(p,x^0)/p^2 .$$

Should the photon have a small mass, the apparent absence of a longitudinal component implies a 4-divergence of the current density vastly smaller than m j_{tr}. Chibisov gives the smallest upper limit of the mass (by far) of 3 x 10^{-27} eV/c^2. It is based on observations of the interstellar gas in the small Magellanic Cloud (Chibisov 1976). Davis' upper limit of 6 x 10^{-16} eV/c^2, based on Jupiter's magnetic field, has a confidence limit of 99.7% (Davis 1975). Watching for anomalous torques in the motion of a Cavendish balance, Lakes very conservatively lowered the limit to half that (Lakes 1998).

5.2 TRANSFORMATION PROPERTY OF THE ELECTROMAGNETIC RADIATION POTENTIAL

The 4-vector nature of the massive potential is quite apparent, be it the solution of the homogeneous or inhomogeneous equation. So, let us look at the massless potential, specifically A_{+1} and A_{-1} ; see eq. (108). The straightforward application of a Lorentz transformation to A_s

$$A_s^\mu(x^\omega) \rightarrow l^\mu{}_\nu A_s^\nu(x^\omega)\,, \quad s = \pm 1 \tag{126a}$$

amounts to transforming the polarization vector $\wp_s(p^\omega)$

$$\wp_s^\mu(p^\omega) \rightarrow l^\mu{}_\nu \wp_s^\nu(p^\omega) \tag{126b}$$

*Eqs. (120b) and (125b) differ in content somewhat from the corresponding eqs. (92) and (98) of Feldman (Feldman 1977).

The right side may be rewritten as

$$l^{\mu}{}_{\nu}H^{\nu}{}_{\omega}(p^{\kappa})\rho^{\omega}_{s} = H^{\mu}{}_{\nu}(l^{\kappa}{}_{\lambda}p^{\lambda})E^{\nu}{}_{\omega}\rho^{\omega}_{s} \tag{127}$$

where H is the 4-by-4 helicity preserving transformation of eq. (99), and E an E(2) transformation as given by eq. (49), specifically the matrix

$$E = H^{-1}(l^{\kappa}{}_{\lambda}p^{\lambda})lH(p^{\kappa}) \ . \tag{128}$$

Since

$$E^{\mu}{}_{\nu}\rho^{\nu}_{\pm 1} = \exp(\pm i\varphi)\rho^{\mu}_{\pm 1} - (1/\surd 2)(\eta_1 \pm i\eta_2)u^{\mu} \tag{129}$$

where

$$u^1 = u^2 = 0, \ u^3 = u^0 = 1$$

we obtain

$$l^{\mu}{}_{\omega}\rho^{\omega}_{\pm 1}(p^{\nu}) = \exp(\pm i\varphi)\rho^{\mu}_{\pm 1}(l^{\nu}{}_{\kappa}p^{\kappa}) - (1/\surd 2)(\eta_1 \pm i\eta_2)v'^{\mu} \tag{130}$$

where

$$v'^{\mu} = H^{\mu}{}_{\nu}(l^{\kappa}{}_{\lambda}p^{\lambda})u^{\nu} \ .$$

Instead of v'^{μ} we can use p'^{μ}/p_i, p_i standing for initial, or reference momentum; see eq. (50b).

The helicity-preserving transformations, see eqs. (50) and (75), used throughout begin with a boost along the third axis. H is such a transformation in the representation (1/2,1/2). Consequently, the zero-components of the polarization vectors, $\rho^{\mu}_{\pm 1}(p^{\nu})$, are zero. The zero-component of the second term on the right of eq. (130) may therefore be expressed in more familiar fashion as $l^{0}{}_{\omega}\rho^{\omega}_{\pm 1}(p^{\nu})$

(Weinberg 1965), resulting in the general component

$$-(1/\sqrt{2})(\eta_1 \pm i\eta_2)p'^{\mu}/p_i = l^0{}_\nu \wp^{\nu}_{\pm1}(p^{\omega})p'^{\mu}/p'^0 . \tag{131}$$

Returning to the transformation of A_s:

$$\begin{aligned} l^{\mu}{}_{\omega}A^{\omega}_{\pm1}(x^{\nu}) = (2\pi)^{-3/2}[\textstyle\int \exp(\pm i\,\varphi(l,p^{\kappa}))A'^{\mu}_{\pm1}(p^{\omega},x^{\nu})d^3p \\ - i\,l^{\mu}{}_{\omega}l^0{}_{\lambda}\,\partial/\partial x_{\omega} \textstyle\int A^{\lambda}_{\pm1}(p^{\alpha},x^{\nu})d^3p/l^0{}_{\kappa}p^{\kappa}] \end{aligned} \tag{132}$$

$A^{\mu}_{\pm1}(p^{\omega},x^{\nu})$ standing for the integrand of the potentials $A^{\mu}_{\pm1}(x^{\nu})$. The prime on the $A_{\pm1}$ signifies a polarization vector with transformed argument; see eq. (130). The phase φ depends on the transformation l and on p^{ω} according to eqs. (128) and (49). So, going to a different Lorentz frame, as distinguished from performing a mere coordinate rotation, produces in addition gradient terms. Compare this to the behavior of the fields $w^{k}_{\pm1}(x^{\mu})$, which have, apart from the important factor $\exp(\alpha)$, the same polarization vectors as the potentials. A Fourier component of a helical 3-vector field undergoing an E(2) transformation is simply multiplied by a phase factor. The radiation potentials undergo E(2) transformations belonging to (1/2,1/2). Since they are transverse, they transform according to the above equation [A.7].

The fields also transform as antisymmetric tensors in 4-space, and thus the gradient terms in the transformed potentials manifestly do not affect them.

5.3 REMARKS ABOUT HELICITY

In the formalism developed for particles without mass we encounter subgroups of SL(2,c) homomorphic (see section 3.4) or isomorphic to the Euclidian group in two dimensions, E(2), the two to one correspondence for spinor, the one to one correspondence for complex vector and 4-vector transformations. We frequently referred to all three representations simply as E(2) transformations.

Except when it came to the potentials A_{+1} and A_{-1}, the effect of the E(2) transformations was always a mere multiplication by a phase factor. Why? Because in the subspaces of each of the representations (1/2,0), (0,1/2), (1,0) and (0,1) we associated only one spin function with a particular momentum direction; see sections 3.2 and 4.1. In the spin-one case the potentials arose as an afterthought, necessitated by gauge invariance. They are coupled to currents.

Say in the case (0,1/2) we also allow a down-spin to go with the up-momentum. While maintaining the equality

$$p^3 = p^0$$

an E(2) transformation will then in general convert the spin-down into a superposition of spin-down and spin-up. If we generalize

$$\sigma_3 p_3 \rightarrow \boldsymbol{\sigma}\cdot\mathbf{p}$$

we obtain

$$(\boldsymbol{\sigma}\cdot\mathbf{p} - p^0)\xi_B = -\kappa'\chi_A\,,\ \xi_B = \xi_{Bp} + \xi_{Ba} \tag{133a}$$

and

$$(\boldsymbol{\sigma}\cdot\mathbf{p} + p^0)\chi_A = 0\,; \tag{133b}$$

p means "parallel", a "anti-parallel". One might say, the lefthanded χ_A is obtained from the "potential" ξ_B. It is "gauge invariant" in the sense that changes in ξ_{Bp} do not affect χ_A (Han 1982, 1986, Ahrens 1987, 1989*). The analogy to the case of the 3-vector w_A and the potential A_{-1} is apparent: the longitudinal components produced by a Lorentz transformation of A_{-1} do not affect w_A.

Next, what about allowing the subspace of, say, (1,0) to contain a spin state or states in addition to the one antiparallel to the momentum? We should then be faced with the equations

$$(\mathbf{J}\cdot\mathbf{p} + p^0)v_A = k\Phi_I\,,\ \mathbf{p}\cdot\mathbf{v}_A = k\Phi_I^0 \tag{134a}$$

*On p. 125 of the latter, $\sigma^3 p$ should have $\sigma^0 p_0$ added to it, and the lower right phase factor in the matrix t should be exp(-iφ).

and

$$(-\mathbf{J}\cdot\mathbf{p} + p^0)\Phi_I - P\Phi_I^0 = 0 \ . \qquad (134b)$$

P is the momentum column vector; see eq, (86). - Equating the left side of eq. (134b) to $k'v_A$ takes us back to massive particles. - For $p^3 = p^0$ eq. (134b) yields $\Phi_I^3 = \Phi_I^0$ and zero for the other components of Φ: it describes massless particles of helicity zero. The v_A play the role of potential.

Evidently the extension of the considered subspaces, plus an extension of those belonging to the representations (1/2,0) and (0,1), leads to potentials, or potential-like quantities. It does not lead to nonhelical massless particles.

Any three-dimensional rotation can obviously be diagonalized. An arbitrary E(2) transformation is more complicated. It involves two very different types of transformations, a rotation about a given axis and a boost along and simultaneous rotation about each of the other two axes. The latter behave mathematically like translations. Examples we encountered, see eqs. (48) and (74), can obviously not be diagonalized, and so cannot represent observables. But one deduces from the commutation relations (47) that a continuum of states allows eigenvalues for the generators E_1 and E_2 (Wigner 1939, 1963):

$$E_1(1 + i\varepsilon J_3)f = E'_1\,(1 + i\varepsilon J_3)f - i\varepsilon E'_2 f \qquad (135a)$$

$$E_2(1 + i\varepsilon J_3)f = E'_2(1 + i\varepsilon J_3)f + i\varepsilon E'_1 f \qquad (135b)$$

f being an eigenfunction of E_1 and E_1 with eigenvalues E'_2 and E'_2, respectively. Consequently,

$$(E_1 + iE_2)(1 + i\varepsilon J_3)f = (1 - i\varepsilon)(E'_1 + iE'_2)(1 + i\varepsilon J_3)f \ . \qquad (136)$$

We may then go to finite rotations,

$$1 + i\varepsilon J_3 \rightarrow \exp(i\varphi J_3) \ .$$

Thusfar, there appears to be no physical evidence for observables corresponding to E_1 and E_2 (Weinberg 1965).

5.4 A RELATION BETWEEN THE ROTATION AND EUCLIDIAN GROUPS

The sets of spinor and of vector equations obtained above obviously also result from the relations between subspaces of L_n, eqs. (23) and (82) plus (86), respectively, if one equates one constant in a set to zero, i.e., if one lets one of the two spinors be helical and likewise one of the 4-vectors Φ. The equations remain relativistically covariant (Ahrens 1989). We know that choosing helical v_A and v_B, called w_A and w_B, respectively, and non-helical Φ results in the homogeneous Maxwell equations (see eq. (76)), and in the fields in terms of the potentials.

As to the behavior of the solutions when κ or κ' and k or k' $\rightarrow$ 0, consider as an example the massive vector-potential. From the integrand in eq. (103) select the column vector that goes with a plane wave of positive frequency and momentum along the third axis:

$$\Phi^{\mu}(p^{\nu}) = \begin{pmatrix} (a^{+}(p) + a^{-}(p))/\sqrt{2} \\ i(a^{+}(p) - a^{-}(p))/\sqrt{2} \\ p^{0} a^{o}/m \\ p a^{o}/m \end{pmatrix} . \tag{137}$$

As viewed by the observer, a positive rotation about the second axis in the frame of the particle is

$$R(\alpha,\theta) = B(\alpha) R(\theta) B^{-1}(\alpha) \ . \tag{138a}$$

In the limit

$$\tanh\alpha = 1$$

this rotation becomes (Ahrens 1984, Han 1984)

$$E(-\eta_1,0,0) = \begin{pmatrix} 1 & 0 & -\eta_1 & \eta_1 \\ 0 & 1 & 0 & 0 \\ \eta_1 & 0 & 1-(\eta_1)^2/2 & (\eta_1)^2/2 \\ \eta_1 & 0 & -(\eta_1)^2/2 & 1+(\eta_1)^2/2 \end{pmatrix} . \tag{138b}$$

A comparison with eq. (49) shows that the generator of this transformation is E_1, and that

$$\eta_1 = \lim(\exp\alpha)\theta/2$$
$$\exp\alpha \to \infty,\ \theta \to 0\ .$$

With exp(α) approaching infinity, the first two components of $\Phi^\mu(p^\nu)$ disappear relative to the last two. This was to be expected from eqs. (86). Indeed,

$$\partial\Phi^3/\partial x_0 - \partial\Phi^0/\partial x_3 = i\,a^{\,o}((p^0)^2 - p^2)/m\ \exp(ip^\mu x_\mu) \tag{139}$$

has a limit of zero.

The helicity-zero components Φ^3 and Φ^0 obviously transform under E(2) like p^3 and p^0: they don't change. The v_A of eq. (134a) however does. It is transformed nontrivially by the matrix of eq. (74b) and the rotation about the third axis. But this turns out to be a gauge transformation: the Φ^3 and Φ^0 of eqs. (134) are not affected. This is the converse of the situation described under 5.2.

It should be remembered that a capital index zero refers to a time-like component, while a lower case index zero implies the third component of a vector in spherical coordinates.

We conclude this discussion with the remark that equating both constants k and k′ to minus the mass and, rather than going over to the (second order) Proca equation, forming the direct sum of eqs. (92b) and (93a), after setting $\Phi_P = 0$, results in the (first order) Kemmer equation (Kemmer 1939, Corson 1953). Here we encounter the connection of the subspaces (1,0) $\oplus$ (0,1) and (1/2,1/2). By considering the six components F^{0k} and F^{lm}, and the four components Φ^μ as the ten components of a vector ψ^k, the Kemmer equation is usually written

$$(\beta_\mu \partial/\partial x_\mu + m)\psi = 0$$

the ten-dimensional β-matrices satisfying the interesting commutation rules

$$\beta_\mu\beta_\nu\beta_w + \beta_w\beta_\nu\beta_\mu = \beta_\mu\delta_{\nu w} + \beta_w\delta_{\mu\nu}$$

(See also Duffin 1938).

5.5 GENERATING MASS

The relativistic Schrödinger or Klein-Gordon equation

$$\begin{aligned} \Box\Phi &= m^2\Phi \\ \Box &= \nabla^2 - \partial^2/(\partial x^0)^2 \end{aligned} \tag{140a}$$

is, for complex Φ, associated with the current

$$j_\mu = -iq(\Phi^*\partial\Phi/\partial x^\mu - \Phi\partial\Phi^*/\partial x^\mu) \tag{140b}$$

which may be obtained from the equation by the same procedure one uses to obtain the nonrelativistic current from the Schrödinger equation. Clearly, equation and current are invariant under gauge transformations of the first kind or "global gauge transformations."

An interaction of the scalar field Φ ("field" used in the general sense, not as distinct from potential) with a massless 4-vector field is introduced by the substitution

$$\partial/\partial x^\mu \rightarrow \partial/\partial x^\mu - iqV_\mu\ . \tag{141a}$$

See the beginning of section 5. This augments the current of eq. (140b) by the term

$$-2q^2\Phi^*\Phi V_\mu \,. \tag{141b}$$

Equation and current are then invariant under local gauge transformations.

Now consider the gauge invariant equation (Goldstone 1961)

$$\Box\Phi - \kappa^2\Phi - 2\lambda\Phi\Phi^*\Phi = 0 \tag{142}$$

which has quite possibly led to more follow-up-work than any other equation in particle physics since the formulation of the Dirac equation. Obtained as the minimum of the "potential" $\kappa^2\Phi\Phi^* + \lambda\Phi^2\Phi^{*2}$, the ground-state-function is

$$\Phi_g = \sqrt{(-\kappa^2/2\lambda)}\,\exp(i\varphi) \tag{143}$$

and not zero, which it would be without the third power term. - In quantum field theory one says that the "vacuum-expectation-value" is given by the above expression, disregarding quantum fluctuations. -

If we now destroy the global gauge invariance or gauge invariance of the first kind by fixing the phase of the ground state, say at zero, express the field as

$$\Phi = (\Phi^1 + i\Phi^2)/\sqrt{2} \tag{144a}$$

and expand Φ^1

$$\Phi^1 = v + \Phi^{1\prime} \tag{144b}$$

where

$$v = \sqrt{(-\kappa^2/\lambda)} \tag{144c}$$

the "Goldstone equation", eq. (142), becomes

$$\begin{aligned} &\Box\Phi^{1\prime} + 2\kappa^2\Phi^{1\prime} - 3\lambda v\,(\Phi^{1\prime})^2 - \lambda(\Phi^{1\prime})^3 - \lambda\Phi^{1\prime}(\Phi^2)^2 = 0 \\ &\Box\Phi^2 - 2\lambda\,v\,\Phi^{1\prime}\Phi^2 - \lambda(\Phi^{1\prime})^2\Phi^2 - \lambda(\Phi^2)^3 = 0\ . \end{aligned} \tag{145}$$

The first equation has a mass term with

$$m = \sqrt{(-2\kappa^2)}\ . \tag{146}$$

The second equation has no mass term. The quanta of Φ^2 are the "Goldstone bosons.". We quote Goldstone: "whenever the original Lagrangian has a continuous symmetry group, the new solutions have a reduced symmetry and contain massless bosons." Depending on the coefficients, the interactions can be treated by perturbation theory.

In hindsight, the next logical step is the substitution (141a) in the Goldstone equation. But instead of looking at the resulting equations for $\Phi^{1\prime}$ and Φ^2, we look at the corresponding equations for V_μ,

$$\Box V_\mu - \partial^2 V_\nu/\partial x^\mu \partial x_\nu = -j_\mu \tag{147}$$

where the current is given by eq. (140b) plus the term (141b). From basic electromagnetic theory we are familiar with the invariance of the left side under

$$V_\mu \to V_\mu + (1/q)\partial\varphi/\partial x^\mu \tag{148}$$

and from our discussion of the current we know that it is invariant under the combined transformations (112a) and (148), replacing in the former equation ψ by Φ. Thus the equation for V_μ is invariant under local gauge transformations. With a choice of $\varphi(x^\mu)$ which makes $\Phi(x^\mu)$ real, V_μ satisfies

$$(\Box - M^2)V_\mu - \partial^2 V_\nu/\partial x^\mu \partial x_\nu = q^2 V_\mu \Phi^{1\prime}(\Phi^{1\prime} + 2v) \tag{149a}$$

once $\Phi(x^\mu)$ has been substituted according to eqs. (144). The new mass of the "vector bosons" is

$$M = q\nu \ . \tag{149b}$$

Now the vector field has three degrees of freedom, as it must, but the total number of degrees of freedom is still four (Englert 1964, Higgs 1964). To accomplish this, the choice of a local gauge which makes the complex field $\Phi(x^\mu)$ real, while convenient, is not necessary. What is, is to make the expansion in the complex plane Φ^1, Φ^2 about the ground state proceed along the line containing origin and ground state. Keeping a field component tangential to the circle of radius $\nu/\sqrt{2}$ at the chosen ground state leads to a total of five degrees of freedom, one of which is clearly unphysical. - As is easily checked, a local gauge transformation cannot lead to a real constant (the ground state) and a purely imaginary field. Thus, only the Goldstone boson can be eliminated. - The new $\Phi(x^\mu)$ is the "Higgs field."

While the work described in previous sections is well founded, based as it is on Lorentz-representation theory and the gauge invariance prompted by quantum mechanics and electromagnetic theory, the work in the present section is rather ad hoc. We bring it up because of its apparently essential role in the remarkably successful electroweak theory. We shall also describe the electroweak theory, though rather briefly, with emphasis on transformation properties and some of the actual transformations, which need to be performed in its development.

5.6 ISOSPIN ET CETERA

Towards the end of section 3.1.4 the interaction of electron and neutrino with the two charged W-particles was mentioned; see eq. (37). Considered initially as the zero-mass quanta of a gauge field, like the photon, the W particles pick up mass from an interaction with the "Higgs field," while the photon does not. A third, neutral, W is added, enabling one to couple three field components to three current components in an invariant manner. All of these components are in "isospace." The two charged and one neutral W-fields may be expressed in Cartesian components in isospace as follows,

$$W^+ = -(W^1 - iW^2)/\sqrt{2}, W^o = W^3, W^- = (W^1 + iW^2)/\sqrt{2} \tag{150}$$

in complete analogy to the spin-up, "sideways" and down fields in terms of their

Cartesian components in ordinary space. In the same way, as we shall see, are "charged and neutral currents" (really current components) related to Cartesian current-components in isospace. The above mentioned coupling may therefore be expressed as $j_\mu W^k_\mu$, an invariant under three-dimensional rotations in isospace. So as to focus on the newly introduced isospace, we omitted the space-time index μ. The proper description of currents and fields generally requires two kinds of indices, each kind referring to a different space; e.g., W^k_μ.

A few words about the concept of a current (actually current density) are in order. According to the usual quantum mechanical definition, the same functions sandwich the vector operator, be it a gradient or a Dirac gamma. But our interest is in transition amplitudes, and the functions in the currentlike part of such an amplitude are going to be different. Nevertheless, we shall continue to refer to that part as current, having already done so under 3), section 3.1.4.

Before proceeding to the discussion of the currents and fields, also some comments on isospace: in his early work on nuclear forces, Heisenberg introduced what became known as an "isotopic spin" to describe neutron and proton as two particles differing only by the plus one and minus one value of a "fifth quantity" (in addition to position and spin). This is in analogy to the up and down spins of, say, the proton (Heisenberg 1932). The small mass difference between neutron and proton is then blamed on the electromagnetic force, to be neglected in the description of nuclear forces. The emission or absorption of a charged meson converts one nucleon state into the other (Yukawa 1935). (If photon emission and absorption could not convert an "up-electron" into a "down-electron" and vice versa, those two states would be two particles. With a permanent B-field, they would be permanently nondegenerate.) Equal probabilities for the emission (and absorption) of negatively and positively charged mesons result in the "charge symmetry" of nuclear forces. Operating on the nucleon doublet $N = \begin{pmatrix} n \\ p \end{pmatrix}$ with the matrix

$\begin{pmatrix} 0 & 0 \\ 1 & 0 \end{pmatrix}$ converts a neutron into a proton, while $\begin{pmatrix} 0 & 1 \\ 0 & 0 \end{pmatrix}$ converts a proton into a neutron. These matrices are usually called $\tau^- = (\tau_1 - i\,\tau_2)/2$ and $\tau^+ = (\tau_1 + i\,\tau_2)/2$, respectively. The τ_k have the form of the Pauli sigmas, but operate in isospace. Kemmer widened the picture by including neutral mesons, doing so in a way which lead to the "charge independence" of nuclear forces (Kemmer 1938, Wentzel 1949). Adhering to the present day convention of assigning isospin + ½ to the proton and - ½ to the neutron, the basic quantum mechanical amplitude is

$$a = g(\bar{p}\gamma_5 n\pi^+ + (\bar{p}\gamma_5 p - \bar{n}\gamma_5 n)\pi^0/\sqrt{2} + \bar{n}\gamma_5 p\pi^-)\ .$$

The three mesons have isospin one. Here we have accounted for the pseudoscalar nature (see section 3.1.5) of the first meson to be discovered, the pi meson, although this property of the "pion" was not established until the mid fifties. In Cartesian coordinates (in isospace) the amplitude is

$$a = g/\sqrt{2}\ (\bar{N}\gamma_5 \boldsymbol{\tau} N \cdot \boldsymbol{\pi}) \ .$$

Spherical and Cartesian components of pions are related like those of the vector bosons; see eq. (150). - Amplitudes are understood to involve a 3-space integration, which is simply denoted by parentheses.

Because the interaction is a scalar in isospace, the isospin, I, is conserved. Nuclear forces so described may depend on the isospin, but not on its third component. While a system of nucleons obeys the Pauli exclusion principle, i.e., the wave function made up of space, spin and isospin parts is antisymmetric under the exchange of any two nucleons, a pion-nucleon system does not, at least not as long as one disregards any possible structure of nucleons and pions. Therefore, pion-nucleon scattering provides a significant test of the meaningfulness of isospin. - Charge independence is observed. But from measurements on nucleon systems alone any observed difference in forces for different I does not automatically imply that isospin is responsible: one must consider the whole wavefunction, which is subject to the exclusion principle. - The state π^+p is pure I = 3/2, while π^-p and π^on are superpositions of I = 3/2 and I = 1/2 states (Brueckner 1952).

In eqs. (77) and (81) we added angular momenta one-half and one to obtain an angular momentum one-half. The examples involved subspaces of the representations (½, 0), (0, ½), (1, 0), (0, 1) and (½, ½) of L_h. For the pion-nucleon states in three-dimensional isospace we thus have

$$-\sqrt{(2/3)}\, p\pi^- + 1/\sqrt{3}\, n\pi^o = (N\pi)^{-1/2}_{1/2} \ .$$

From orthogonality between I = 1/2 and 3/2 states, we then have

$$\sqrt{(2/3)}\, n\pi^o + 1/\sqrt{3}\, p\pi^- = (N\pi)^{-1/2}_{3/2}$$

with presumably obvious notation. From solving these equations for the scattering states we deduce

$$\sigma(\pi^+ p \to \pi^+ p) : \sigma(\pi^- p \to \pi^- p) : \sigma(\pi^- p \to \pi^0 n)$$

$$= |a(3/2 \to 3/2)|^2 : |a(3/2 \to 3/2) + 2a(1/2 \to 1/2)|^2/9$$

$$: 2|a(3/2 \to 3/2) - a(1/2 \to 1/2)|^2/9 \ .$$

Indeed, the $a(3/2 \to 3/2)$ behaves quite differently from the $a(1/2 \to 1/2)$ amplitude. It exhibits a resonance of 120 MeV width at half maximum about a peak at 195 MeV pion kinetic energy. The resonance was first observed in 1951 by Fermi and coworkers (Anderson 1952). - In the energy range of this resonance we therefore have the famous cross section ratios 9 : 1 : 2. The resonance really has four components, called Δ^{++}, Δ^{+}, Δ^{0}, and Δ^{-}. They form a quartet of isospin states, which in turn is part of the well known baryon decuplet of spin 3/2 and parity +1, completed with the discovery of the Ω^-, which is an isosinglet (Barnes 1964). Between the quartet and the singlet there are a triplet and a doublet [A.11].

These facts found an explanation at a deeper level than that of nucleon, pion and delta isospins, the quark level (Gell-Mann 1964, Zweig 1964). Accordingly, the resonance is made up of three quarks, just like the nucleons, but with spin and isospin 3/2 rather than 1/2. For both kinds of particles the customary three-part wave function, see above, is symmetric, since the space part is assumed to be made up of s-states. "Quark color" came to the rescue (Greenberg 1964). Each quark can appear in any of three "colors", which are commonly chosen to be the primary colors. Since color never makes its appearance, the color wave function is a singlet, which is antisymmetric. At "low energy" (as far as high energy physics goes) the four quarks and the antiquark (mesons are explained as quark-antiquark compounds) in the pion-nucleon scattering manage to form a "3/2, $3/2^+$" but not a "1/2, $3/2^+$" three-quark resonance. (For I = 1/2, the isospin wave function of the three-quark system is antisymmetric.)

Even before the introduction of quarks, the isomultiplets of "normal particles" were joined by those of "strange particles". In their data analysis, Gell-Mann and Nishijima (Gell-Mann 1953, Nishijima 1954) independently arrived at a relation between electric charge (in units of the positive electronic charge), third I-spin component, "baryon number" and "strangeness",

$$Q = I_3 + \frac{B+S}{2} \ .$$

For baryons and antibaryons B is 1 and -1, respectively, while it is zero for mesons. See also section 6. Produced strongly together with another strange particle, the so called "associated production" (Pais 1952), a strange particle decays weakly. In a strong reaction S is conserved, not so in a weak decay. A new quantum number was born. The sum B+S is called the hypercharge, Y. Consider one isomultiplet of strange baryons, also called "hyperons", the triplet $\Sigma^+, \Sigma^0, \Sigma^-$ of spin-1/2 particles. It has the quantum numbers I = 1, B = 1 and S = -1. In a sufficiently energetic π^- - p collision Σ^- and K^+ are produced in association, K^+ being a (strange) meson with S = +1. As predicted, it turned out to be one component of an isodoublet, the other component being the K^0 meson.* The latter can, e.g., be produced in the reaction $\pi^- + p \rightarrow \Lambda + K^o$, Λ being a singlet hyperon. But what about the later interpretation in terms of quarks? Σ^-, e.g., has two down and one strange quark, the latter with $I_3 = 0$ and S = -1. (The reader has presumably heard enough about quarks, to know of the fractional electrical charges + 2/3 and - 1/3). B is taken to be 1/3, since three quarks make a baryon. From an interpretation of the mass differences between three-quark isomultiplets in terms of interquark spin-spin forces, the strange quark is 175 MeV/c^2 heavier than the (almost equal) up and down quarks (Perkins 1987). Neighboring isomultiplets within the same "SU(3) multiplet" (Gell-Mann 1962(1), Ne'eman 1961), e.g., the baryon octet composed of the nucleon doublet, the sigma triplet, the lambda singlet and the Xi doublet, are separated by mass differences of roughly two hundred MeV [A.11]. The mass differences within an isomultiplet, on the other hand, are of MeV order. Since isomultiplets transform according to the representations of the group SU(2), and since SU(3), the "special unitary group in three dimensions" (Fronsdal 1963), contains an SU(2) subgroup, isospace is consequently the one to pick for subspace of that subgroup. Isomultiplets of the same I are distinguished by their S. In the octet case p, n have S = 0, Ξ^o, Ξ^- have S = -2. There are three ways to pick SU(2) subspaces [A.11]. But if considered as members of either of the other two kinds of SU(2) multiplets, the states within one such multiplet will exhibit mass differences of about 200 MeV. Clearly, isomultiplets make more sense.

Just as proton, neutron or up quark, down quark form a subspace of the "fundamental representation" of SU(2), so do up quark, down quark and strange quark form a subspace of the fundamental representation of SU(3). The eight-by-eight representation, which has the baryon octet as a subspace, is contained in the direct product of three fundamental representations of SU(3). So is the ten-by-ten representation, which has the mentioned baryon decuplet as a subspace (Perkins 1987). There the Ω^- interpreted in terms of three strange quarks, demonstrates

*At the time the assignment I = 1 to an isotriplet K^+, K^o, K^- would have been more popular. But, indeed, K^- was found to have a partner, $\bar{K}^o$.

most obviously the need for color. - It should be kept in mind, that the prediction of the Ω^- (Gell-Mann 1962(2), Ne'eman 1962) preceded Gell-Mann's quark and Zweig's "aces" hypotheses.

Working one's way up to progressively heavier families of baryons and mesons, the strangeness story repeats itself. One adds c (charmed), b (bottom) and t (top) quarks. The distinguishing quantum numbers are collectively referred to as "flavor".*

It is worth noting that the positively charged quarks are heavier than their negatively charged partners, with the exception of q_u and q_d. One would expect the proton to be heavier than the neutron, if electromagnetism were solely responsible for their mass difference. But presumably the electromagnetic effect is outdone by the mass of the down quark, which appears to be bigger than that of the up quark.

Through all this one needs to keep in mind, that the elementary strong force is the color force. Like the pion-nucleon force it is charge independent, but most simply so. To achieve charge independence, the coupling constants for the emission and absorption of pions had to be arranged so that electrically neutral pions were coupled to p p - n n with $1/\sqrt{2}$ the strength of the equal couplings of the two electrically charged pions to the charge exchanging nucleons. This was (and is) described by the scalar product of two isovectors. The strength of these couplings has nothing to do with the electronic charge. - The quanta which the quarks are thought to emit and absorb strongly, the "gluons", not only carry the charges of the charge exchange between emitting and absorbing quarks, but their coupling to the quarks is given by the same charges, not some other coupling constant. This is the situation in a (non-Abelian) "gauge theory".

There are eight gluons, forming an irreducible subspace of the group SU(3), not to be confused with the SU(3) of isospin and strangeness (or hyper-charge). As we shall see in the discussion of the electroweak theory, the situation here is at once more complicated and simpler than there. The group SU(3) has eight generators of transformations, the group of the electroweak theory, U(1) $\otimes$ SU(2), has four. But the gluons are assumed to remain massless. Except for the photon, the quanta of the electroweak interaction, the vector bosons, acquire mass. - In the color force, no singlet gluon is supposed to participate.

Cini and Gamba extended the isotopic spin concept to electrons, muons and neutrinos, already associating a neutrino ("neutretto") with the muon different from the neutrino associated with the electron (Cini 1953). Postulating charge independence, they obtained what would come to be called "weakly interacting

* For colored pictures of multiplets including strangeness and charm, see Glashow 1975 and Schwitters 1977).

neutral currents". These were supposed to give rise to weak processes such as

$$e^+ + e^- \rightarrow \mu^+ + \mu^-$$

and presumably, although not explicitly mentioned by the authors, such as

$$\nu + p \rightarrow p + \nu$$

- A weakly interacting neutral current had been suggested many years before in work involving β decay (Oppenheimer 1939). -

The electron-neutrino electron and muon-neutrino muon doublets are referred to as "generations", and are joined by the tau-neutrino tau doublet. The analogous quark generations are the up and down, the charmed and strange and the top and bottom quarks. The associated isospin is referred to as "weak isospin". Henceforth, the isospin of interest will be this "weak isospin" (see also section 7.2.1).

Chapter 6

ELECTROWEAK THEORY

The idea of charge independence of the weak interaction is maintained, and, based on experiment, a distinct neutrino is associated with each charged lepton, nowadays e, μ (Danby 1962) and τ. But, the electroweak theory is built on local gauge invariance, and an interaction of currents with 4-vector fields, not on any current-current interactions. Thus the original Fermi and Gamow-Teller interactions are actually of second order.

In the case of electron and its neutrino one writes (a for amplitude)

$$a = -(j_k^\mu W_\mu^k),\ j_k^\mu = g_l/2\ \bar{l}_L \tau_k \gamma^\mu l_L \tag{151a}$$

where

$$l_L = \begin{pmatrix} \nu \\ e \end{pmatrix}_L \tag{151b}$$

L meaning left handed. For the electron

$$e_L = (1-\gamma_5)/2\ e,\ e \equiv \psi_e(x^\nu) \tag{152}$$

and similarly for the neutrino. In the representations diagonal in γ_5 which we discussed, the left-handed components are the lower components of the 4-spinor. It must be remembered that in the case of massive particles handedness is not the same as helicity. The most useful form of the amplitude is

$$a = -g_l/\sqrt{2}\ (\bar{\nu}_L\gamma^\mu e_L W_\mu^{-*} + 1/\sqrt{2}\ (\bar{\nu}_L\gamma^\mu \nu_L - \bar{e}_L\gamma^\mu e_L)\, W_\mu^o + \bar{e}_L\gamma^\mu \nu_L W_\mu^-)) \ .(153)$$

Note the W_μ^{-*}. We know that the spin-up nature (in isospace) of W_μ^+ requires a minus sign in front; see the comments following eq. (104). But then so does the isospin-down current component, which is $-\bar{\nu}_L\, \gamma_\mu\, e_L$, because one has

$$\begin{pmatrix} \nu \\ e \end{pmatrix} \sim \begin{pmatrix} \bar{e} \\ -\bar{\nu} \end{pmatrix} \tag{154}$$

as behooves subspaces of SU(2).

An ingenious step in the construction of the electroweak theory is the identification of the right-handed electron with an isoscalar. This leads to the current

$$j_R^\mu = g_o Y_R/2 \ \bar{e}_R \gamma^\mu e_R \ . \tag{155a}$$

There is a second isoscalar current, formed by isospinors,

$$j_L^\mu = g_o Y_L/2 \ (\bar{\nu}_L \gamma^\mu \nu_L + \bar{e}_L \gamma^\mu e_L) \ . \tag{155b}$$

Compare it to the neutral term on the right of eq. (153).

As for the constants, g_o /2 is the coupling constant of the isoscalar and $g_1/2$ that of the isovector current. The "hyperchange" Y may be viewed as the multiple of an elementary charge of a particle described by l_L (see eq. (151b)) or by

$$l_R = e_R \ . \tag{156}$$

No hypercharge needs to multiply the single isovector current of eqs. (151a).

The 4-vector field, with which the entire isoscalar current

$$j^\mu = j_R^\mu + j_L^\mu \tag{157}$$

interacts, is called B_μ. As in the left of eqs. (151a), the amplitude has a minus sign.

In the electroweak theory there is no theoretical connection between the coupling constants g_0 and g_1.

- Before the work of Goldstone, Higgs and others provided a means of generating mass, Glashow proposed the coupling of four massive vector-bosons to lepton currents, the couplings being invariant under global U(1) (the one-dimensional unitary group) and SU(2) transformations, but not the mass terms. Diagonalization of the mass terms could then result in linear combinations of the two neutrals among the original vector fields, one combination being the massless electromagnetic field. At the time, Glashow used one (Majorana) neutrino with electron and positron (Glashow 1961). -

We saw that from a quantum-mechanical viewpoint the electromagnetic field can be introduced in terms of a 4-potential from the demand that an equation, e.g. the Dirac equation, or the Klein-Gordon equation for a complex field, be invariant under a local gauge transformation; see the beginnings of sections 5 and 5.5. Now we have the potentials $B_\mu(x^\nu)$ and $W_\mu^k(x^\nu)$ to consider. Their presence in the equations for the lepton spinors in the form

$$\gamma^\mu(\partial/\partial x^\mu - ig_0\,Y/2\,B_\mu - ig_1/2\,\boldsymbol{\tau}\cdot\mathbf{W})\,l + ml = 0 \tag{158}$$

gives invariance under

$$l \to (1 + iY/2\;\delta\varphi + i/2\,\boldsymbol{\tau}\cdot\delta\boldsymbol{\varphi})\,l \tag{159a}$$

$$\begin{aligned} \textit{and} \qquad B_\mu &\to B_\mu + 1/g_0\;\partial\delta\varphi/\partial x^\mu \\ W_\mu^k &\to W_\mu^k + 1/g_1\;\partial\delta\varphi^k/\partial x^\mu + iJ^k_{lm}\,\partial\varphi^l\,W^m_\mu\;. \end{aligned} \tag{159b}$$

J^k_{lm} is the k, m element of the l th three-by-three angular-momentum matrix; see eqs. (67). Here **J** operates in isospace, like $\boldsymbol{\tau}$. The use of infinitesimal transformations may be excused by the structure of the second of eqs. (159b). - The third term on the right of the latter is already present, if the transformations are merely global. It should be self-explanatory.

The transformations of the isospinors and isovectors are "non-Abelian," i.e., they are noncommutative. Such local gauge transformations were first used in connection with strong interactions (Yang 1954).

- Some words about leptons, the elementary spin-one-half particles, which do not interact strongly. Nowadays these are e, μ and τ, their neutrinos, and the antiparticles of all of these. Should some of the neutrinos turn out to be Majorana neutrinos, then they are their own antiparticles.

Konopinski and Mahmoud proposed what amounts to "lepton-number conservation," the conservation of leptons minus antileptons. At the time fewer leptons were known, and what constitutes a "normal particle", as opposed to an antiparticle, appeared to be different from what we now know to be the case. But the law was correct. (They called e^- and μ^+ normal.) Together with their "conservation of heaviness," to become "conservation of baryon number," they eliminated all unwanted processes and only these (Konopinski 1953). The currents discussed above comply with lepton-number conservation. In 1938 Stueckelberg had proposed the conservation of nucleons (Stueckelberg 1938). -

We know how the electromagnetic field is coupled to the currents:

$$a = -e(\bar{e}_L\gamma^\mu \bar{e}_L + \bar{e}_R\gamma^\mu e_R)A_\mu \tag{160}$$

the e in front standing for the charge on the electron, the others for the left and right handed parts of the, at this point, zero-mass electron wave-functions.

The real orthogonal matrix

$$1/N\begin{pmatrix} g_1 & +g_o Y_L \\ -g_o Y_L & g_1 \end{pmatrix}, \; N^2 = g_o^2 Y_L^2 + g_1^2 \tag{161}$$

allows us to express the fields B_μ and W_μ^o in terms of A_μ and "Z_μ", so that the transition amplitude (N for neutral)

$$\begin{aligned} a_N = -g_o/2\,(Y_R\bar{e}_R\gamma^\mu e_R + Y_L\,(\bar{\nu}_L\gamma^\mu\nu_L + \bar{e}_L\gamma_\mu e_L))B_\mu \\ - g_1/2\,(\bar{\nu}_L\gamma^\mu\nu_L - \bar{e}_L\gamma^\mu e_L)\,W_\mu^o \end{aligned} \tag{162}$$

becomes

$$a_N = -(g_o g_1/2N)(Y_R \bar{e}_R \gamma^\mu e_R + 2\,Y_L \bar{e}_L \gamma^\mu e_L) A_\mu$$

$$-1/2N\,[(g_o^2 Y_L)\,(Y_R \bar{e}_R \gamma^\mu e_R + Y_L(\bar{\nu}_L \gamma^\mu \nu_L + \bar{e}_L \gamma^\mu e_L)) \qquad (163a)$$

$$+ g_1^2(\bar{\nu}_L \gamma^\mu \nu_L - \bar{e}_L \gamma^\mu e_L)]\,Z_\mu \;.$$

The neutrinos are not coupled to A_μ. Choosing $Y_L = Y_R/2 = -1$ gives the correct electron current, with

$$g_o g_1\, Y_L/N = e \;. \qquad (164)$$

For positive g_0 and g_1 the hypercharge of the left-handed lepton doublet must obviously be negative. Always appearing as a factor of g_0 , its magnitude is usually set equal to one.

$$With \quad g_o/N = \sin\theta \quad and \quad g_1/N = \cos\theta$$

the expression (163a) may be written

$$-(\bar{e}_R \gamma^\mu e_R\, g_o \sin\theta\, (-Y_R/2) + \bar{e}_L \gamma^\mu e_L\, (g_1 \cos\theta\, I_3 - g_o \sin\theta\, Y_L/2)$$

$$+ \bar{\nu}_L \gamma^\mu \nu_L\, (g_1 \cos\theta\, I_3 - g_o \sin\theta\, Y_L/2))\,Z_\mu \;. \qquad (163b)$$

θ is usually called the "Weinberg angle", θ_W . Sometimes it is referred to as the Glashow-Weinberg angle.

Just as they do in the equation for the leptons, eq. (158), will the gauge fields appear in the equation for the Higgs field, Φ (see page 77), because, as proposed by Weinberg and Salam, the pertinent Higgs field is an isodoublet. Once B_μ and W_μ^o are expressed in terms of A_μ and Z_μ , the equation is

$$(\partial/\partial x^{\mu} - iM_{\mu})(\partial/\partial x_{\mu} - iM^{\mu})\Phi - \kappa^2\Phi = 2\lambda\Phi\Phi^*\Phi \tag{165}$$

$$M_{\mu} = \begin{pmatrix} (g_o g_1/2N)(Y+1)A_{\mu} + ((-Yg_o^2+g_1^2)/2N)Z_{\mu} & -g_1/\sqrt{2}\, W_{\mu}^{+} \\ g_1/\sqrt{2}\, W^{-} & (g_o g_1/2N)(Y-1)A_{\mu} - ((Yg_o^2+g_1^2)/2N)\, Z_{\mu} \end{pmatrix}.$$

For the doublet Φ it is customary to choose

$$\Phi = \begin{pmatrix} \Phi^{+} \\ \Phi^{o} \end{pmatrix} \tag{166}$$

and the hypercharge to be plus one. Then the doublet with Φ^- on the bottom gives the antiparticles, whose hypercharge is minus one. For A_μ not to be coupled to the vacuum-expectation-value and thus to remain massless, the neutral Higgs gets the nonzero ground state. - From the viewpoint of quantum field theory, this is also manifestly essential for charge conservation: the field function in the expansion about the vacuum-expectation-value (see statement following eq. (143)) is then an operator and, appearing in transition amplitudes in odd powers, would destroy and create electric charge if it originated from Φ^+. -

By looking back at how the mass term came about in the case of the single 4-vector, V_μ, eqs. (147) and (141b), we can immediately deduce the masses of the W and Z particles from the matrix M,

$$M_W = g_1 v/2 \text{ and } M_Z = Nv/2 \ . \tag{167}$$

The two degrees of freedom of the complex Φ^+ and the Goldstone Boson of Φ^0 were surrendered to make the three vector Bosons massive. This is done with the local gauge transformation $\exp(i\boldsymbol{\tau}\cdot\boldsymbol{\varphi}(x^\mu))$, turning Φ^+ as well as the imaginary part of Φ^0 into zero. Of the invariance under the local transformations in, properly expressed, the direct product of groups U(1) ⊗ SU(2) only an invariance under the transformations

$$\Phi \rightarrow \exp(iQ\varphi(x^\nu))\Phi$$
$$\textit{and}\ A_\mu \rightarrow A_\mu + (1/e)\partial\varphi(x^\nu)/\partial x^\mu \tag{168}$$

persists. Only A_μ remains massless. For the sake of generality, the electric charge in units of e has been inserted. Clearly, see eqs. (165) and (163),

$$Q = \tau_3/2 + Y/2\ . \tag{169}$$

This is not the original U(1) invariance, which involved the field B_μ. It is the invariance leading to the conservation of charge. The Higgs component

$$\Phi^0 = 1/\sqrt{2}\,(v + H) \tag{170}$$

has zero charge. It is unaffected by the gauge transformation (168). To achieve invariance, any other local gauge transformations would have to change Φ^o. But v has been fixed!

The above relation between charge, isospin and hypercharge, first given by Gell-Mann and Nishijima in their systematization of ordinary and strange baryons and mesons, was for weak interactions proposed by Glashow (Glashow 1961).

The Gell-Mann Nishijima relation established the connection between charge and (strong) isospin for ordinary and strange particles, see section 5.6. In the case of ordinary particles it merely pointed out the "charge displacement" of nucleons and its absence for π mesons. Elevating the hypercharge to the status of separate and equal participant in a group structure was first done successfully within global SU(3) as applied to strong (not yet color) interactions. Doing something somewhat similar for the weak hypercharge, Glashow incorporated it in the U(1) part of the group SU(2) ⊗ U(1). In the infinitesimal transformations (159) we see it (presumably not finally) as generator of the local U(1) group, part and parcel of the Salam-Weinberg theory.

In eq. (163b) we see quite clearly that a "neutral current" (this conventional jargon includes currents like $\bar{e}\gamma^\mu e$) is multiplied by

$$g_1\cos\theta\, I_3 - g_o\sin\theta\, Y/2\ .$$

It is customary to rewrite this as

$$g_1/\cos\theta \; (\cos^2\theta \; I_3 \; - \; \sin^2\theta \; Y/2)$$

$$= g_1/\cos\theta \; (I_3 - Q\sin^2\theta) \; .$$

From eq. (164) we know that $g_1 = -e/\sin\theta$.

Let the left-handed neutrino be a Weyl neutrino. This, of course, makes the left-handed electron massless too (see eq. (151)), and, because of the interaction with the electromagnetic field, the right-handed electron also. Not being treatable like the quanta of 4-vector gauge-fields, leptons then get their masses on account of invariant amplitudes put in by hand. E.g.,

$$i\,a = g(\bar{l}_L \Phi e_R + \bar{e}_R \Phi^\dagger l_L) \; . \tag{171}$$

Expanding the Higgs field

$$\Phi \; = \; 1/\sqrt{2} \begin{pmatrix} 0 \\ \mathrm{v} + H \end{pmatrix}$$

(see statement following eq. (166)), one obtains

$$i\,a \; = \; g/\sqrt{2}\, (\bar{e}_L e_R + \bar{e}_R e_L)(\mathrm{v} + H) \;\; ; \tag{172}$$

v is a constant, and therefore the initial and final electron functions multiplying it are the same.

Up to now there was no real need to talk about Hamiltonians or Lagrangians. To easily interpret the last equation, it is indicated to do so. Whether classical or quantum, field equations may usually be derived from Hamiltonians or Lagrangians. If one chooses the latter approach, the field equations are Euler-Lagrange equations,

$$\frac{\partial\mathcal{L}(f_k(x^\nu),\partial f_k/\partial x^w)}{\partial f_k(x^\nu)} - \frac{\partial}{\partial x^\mu}\frac{\partial\mathcal{L}(...)}{\partial\frac{\partial f_k}{\partial x^\mu}} = 0 \ . \tag{173}$$

The only sum is over μ. The 3-space integral over the "Lagrangian density," $\mathcal{L}$, is the Lagrangian, L. Take the case of the free Dirac equation, so that $f_k = \psi_k$, $k = 1$ to 4.

$$-i\mathcal{L} = 1/2\,(\bar{\psi}\gamma^\mu\partial\psi/\partial x^\mu - \partial\bar{\psi}/\partial x^\mu\gamma^\mu\psi) + m\bar{\psi}\psi \tag{174}$$

placed in the Euler-Lagrange equation will result in the Dirac equation. We can express the scalar $\bar{\psi}\psi$ as

$$\bar{\psi}\psi = \bar{\psi}_L\psi_R + \bar{\psi}_R\psi_L \ . \tag{175}$$

Compare to eq. (172)! So we are led to identify

$$gv/\sqrt{2} = m \tag{176}$$

where m is the mass of the electron (Weinberg 1996*). Unfortunately, every lepton mass thus generated is proportional to a new coupling constant. The expression for $\bar{\psi}\psi$ tells us also something else. The Lagrangian density of the electroweak theory, prior to the mass generation by a Higgs field, cannot contain a mass term, at least not as long as one wishes the Lagrangian to be invariant under SU(2) in isospace, of course in addition to being invariant under L_h. For, as we have learned, the assumption of a left-handed isodoublet and a right-handed isosinglet is crucial to the ultimate success of the theory, and a mass term with products of isodoublets and isosinglets is clearly not invariant. Thus the assumption of massless leptons to begin with is an essential part of the electroweak theory.

* There each component of the doublet Φ is larger by the factor √2 than Goldstone's original complex Φ.

Eq. (172) tells us also that there is an electron-electron-Higgs interaction.

We encountered lepton - vector Boson interactions (including the electromagnetic potential), vector Boson - Higgs interactions, and lepton - Higgs interactions. There are also interactions between the vector Bosons, which are best obtained by starting from a Lagrangian density (Aitchison 1989). As one can surmise by looking at eqs. (149), there exist three and four vector-Boson "vertices." Here we shall not delve into those interactions.

We have only dealt with leptonic spin-one-half particles. Quarks also interact electromagnetically and weakly. In this respect they only differ from leptons, apart from their masses, by their fractional charges and the fact that both, the charge-two-thirds (isospin-up) and the charge-one-third (isospin-down) quark have right handed components, which transform as isoscalars. An isoscalar neutrino on the other hand, not coupling to A_μ, would not couple to Z_μ either, because its hypercharge should be zero. It is a kind of "sterile neutrino".

Just as the leptons, the quarks may in their electroweak behavior be described by weak isospin and weak hypercharge, the latter of right-handed and left-handed varieties. Since left-handed quarks, like left-handed leptons, are successfully described as "generations" of isodoublets, their weak isospin is always one-half (quite differently from their strong isospin). Consequently, all left-handed quarks have $Y_L = 1/3$. Right-handed quarks behave like isoscalars. Consequently, right-handed quarks have $Y_R = 4/3$ and $- 2/3$, respectively.

6.1 NEUTRAL CURRENTS

Between 1964 and 1969 several high energy experiments to detect the scattering of "muon neutrinos" (as distinct from the electron neutrinos discussed above) by protons yielded upper limits on the cross section between three and fifty percent of the known cross section for the process neutrino on neutron yielding proton and muon (see also preface). Then in 1973 neutral particle (neutrino) induced events without the appearance of muons or electrons were recorded (Hasert 1973, Benvenuti 1974). Some three years later the neutrino scattering by protons was identified (Lee 1976), and again three years later, this time at low energy, the neutrino-disintegration of the deuteron in the neutral mode (Pasierb 1979).

The massive vector-Bosons were first observed in 1983 (Arnison 1983, Banner 1983, Bagnaia 1983).

6.2 COUPLING CONSTANTS

Eq. (164) provides us with a connection between the isoscalar and isovector coupling constants g_o and g_1 and the charge of the electron. It is nice also to have a relation between g_o and g_1 and the Fermi coupling constant G of β decay, and therefore of low energy weak interactions. We consider the decay of the muon. The amplitude is

$$a = 2\sqrt{(2)}\,G/\cos\theta_C\ \ (\bar{\nu}_{\mu L}\gamma^{\kappa}\mu_L \nu_{\mu R}\,\bar{e}_L\gamma_{\kappa}\nu_{eL})\ . \tag{177}$$

The strange form of the effective coupling constant has historical reasons: G, the Fermi coupling constant, comes from the original formulation of β decay. E.g., the decay of the neutron would just contain G. The increase in strength described by $1/\cos\theta_C$ is a result of the Cabibbo theory, see section 7.2, according to which the charged currents of hadrons are rotated through θ_C. The $\sqrt{2}$ is due to the choice of ν_{eL} , which, as we know, contains a $(1 - \gamma_5)/2$, while the transition probability of ($\bar{\nu}_e$ standing for electron antineutrino)

$$\mu^- \to e^- + \bar{\nu}_e + \nu_\mu \tag{178}$$

is unaffected by the substitution

$$\psi_\nu \to (1 - \gamma_5)/\sqrt{2}\ \psi_\nu\ .$$

In pre-parity-nonconservation days one used ψ_ν and ψ_e without any projection operators. As to the e_L in the above amplitude, it is explained by

$$(1 - \gamma_5)/2\ (1 - \gamma_5)/2 = (1 - \gamma_5)/2\ .$$

Finally, the factor 2 is due to having rewritten the “V-A” expression

$$(G/\cos\theta_C)(\bar{\nu}_\mu\gamma^{\kappa}\mu\,\bar{e}\gamma_{\kappa}\nu_e - \bar{\nu}_\mu i\gamma^{\kappa}\gamma_5\,\mu\,\bar{e}\,i\gamma_{\kappa}\gamma_5\,\nu_e)$$

as the right side of eq. (177) not merely more compactly, but also more meaningfully.

If not before, then certainly since the discovery of the massive vector-Bosons have the weak decays been viewed as second-order processes. Thus the effective coupling-constant in eq. (177) is equated to $(g_1)^2/2M_W^2$, the inverse square of the charged-W mass being the low energy approximation of the "vector-meson-propagator" (Byers 1958). The decay (178) is pictured by the Feynman diagrams of Fig. 2.

From the knowledge of $G/\cos\theta_C$ and M_W we get g_1 , and consequently from e also g_0 . This gives the Weinberg angle, θ_W , by

$$\tan\theta_W = g_0/g_1 \tag{180}$$

and thus the ratio

$$M_W/M_Z = \cos\theta_W \ . \tag{181}$$

Values of the boson masses are M_w = 80.396 (61) GeV/c^2 (WWW Rev. Part. Phys. 1999) and M_z = 91.1871 (21) GeV/c^2 (Swartz 1999).

The Higgs vacuum-expectation-value follows from M_W and g_1, as well as from M_Z , g_0 and g_1 , see eqs. (167). Values of G, e and θ_w are (NIST 1998),

$$G/(\hbar c)^3 = 1.16639\ (1) \times 10^{-5}\ \mathrm{GeV}^{-2}$$

$$e = -1.602176462\ (63) \times 10^{-19}\ \mathrm{C}$$

and $\sin^2\theta_w$ = 0.2224 (19).

For $\cos\theta_C$ see eq. (194).

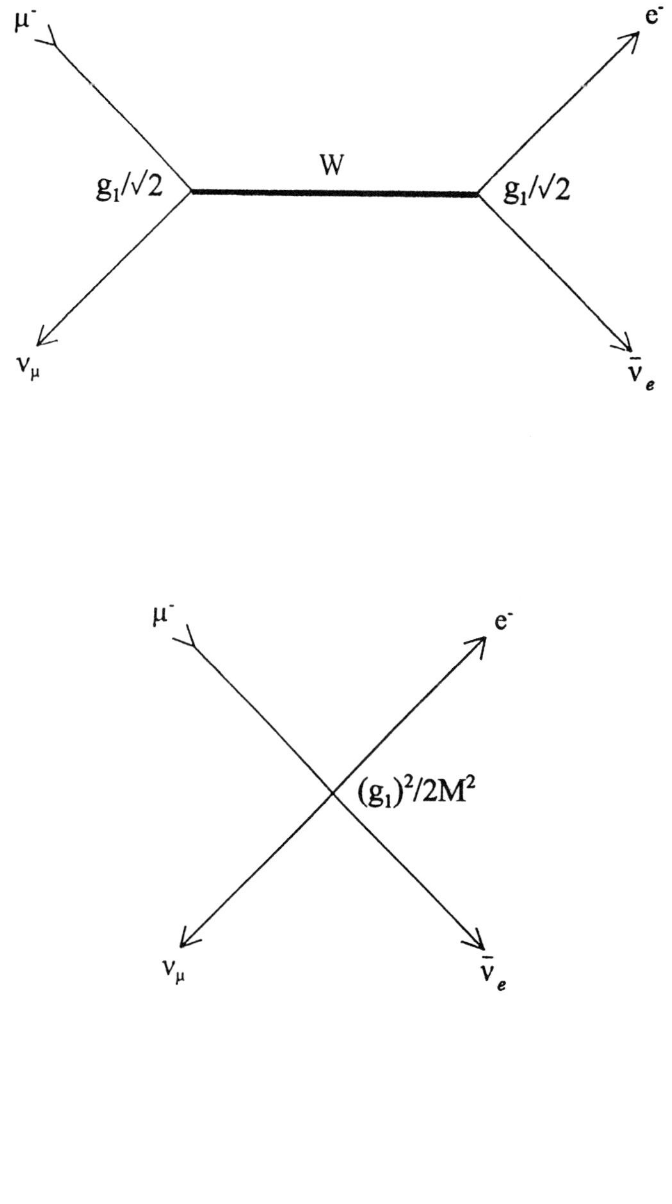

Figure 2: Feynman diagrams for the decay of the muon.

Chapter 7

MASSIVE NEUTRINOS

Accounting for neutrino masses extends particle physics beyond the "standard model", which encompasses the original electroweak theory and quantumchromodynamics.

7.1 TRITIUM DECAY

The "direct method" of looking for the mass of the electron neutrino involves the precision measurement of a beta spectrum near the endpoint energy, usually the spectrum of tritium. The small mass difference between tritium and helium-three, resulting in a beta endpoint energy of about 18.57 keV, and not to forget the half life of 12.33 ± .06 y, which allows considerable luminosities, make this spectrum the prime candidate for determining the mass, if any, of the electron neutrino. The energy dependence of the electron distribution is given by the three-particle phase space factor, multiplied by the Fermi function (Fermi 1934, Boehm 1987, Feynman 1962),

$$P(E) \propto F(Z,E)p(E+m_e c^2)(E_0 - E)\sqrt{((E_0 - E)^2 - m_\nu^2 c^4)} \qquad (182)$$

with E the kinetic energy and p the momentum of the electron. E_0 is the endpoint energy, and Z the atomic number of the daughter nucleus. The last two factors are neutrino energy and momentum. The recoil energy has been neglected. The low Z and the fact that one is dealing with a "superallowed" transition, make for rather reliable calculations.

The "Troitsk" experiment involves an electrostatic spectrometer with a strong longitudinal, nonuniform magnetic field for collimation of the β particles. This field is produced by superconducting coils (Lobashev 1996).

An upper limit on the electron-neutrino mass is

$$m_\nu < 2.5\ eV/c^2$$

with a 95 percent confidence level (Lobashev 1999).

7.2 NEUTRINO OSCILLATIONS AND COHERENCE

If some or all neutrinos have mass, one might expect "neutrino oscillations", (Pontecorvo 1957, 1958, 1967, Gribov 1969). Nondecaying neutral particles created and destroyed courtesy of the weak interaction may not be mass eigenstates. We write the wave function of a mass eigenstate as

$$\nu_k(t) = \nu_k \, exp(-iE_k t) \qquad no\ sum \tag{183}$$

and the wave function of an observed neutrino as a sum over the mass-eigenstates

$$\nu_l(t) = U_{lk}\, \nu_k(t) \qquad l = e, \mu, \tau;\ k = 1-3\ . \tag{184}$$

Solving the last equation for $\nu_k(t)$ at t equal to zero, using this for ν_k in the previous equation and putting the so obtained $\nu_k(t)$ back into the last equation, yields

$$\nu_l(t) = U_{lk}\, U^{-1}_{kl'}\, \nu_{l'}\, \exp(-iE_k t)\ . \tag{185}$$

Restriction to just two kinds of neutrinos, say

$$l, l' = e, \mu;\ k = 1,2$$

gives

$$\begin{aligned} \nu_\mu(t) = {} & \nu_\mu(\cos^2\theta \exp(-iE_2 t) + \sin^2\theta \exp(-iE_1 t)) \\ & + \nu_e \sin\theta \cos\theta\, (\exp(-iE_2 t) - \exp(-iE_1 t)) \end{aligned} \tag{186a}$$

and

$$\begin{aligned} \nu_e(t) = {} & \nu_e(\sin^2\theta \exp(-iE_2 t) + \cos^2\theta \exp(-iE_1 t)) \\ & + \nu_\mu \sin\theta \cos\theta\, (\exp(-iE_2 t) - \exp(-iE_1 t))\ . \end{aligned} \tag{186b}$$

For the "mixing matrix" (U_{lk}) we have used the two-dimensional rotation through angle θ. This is possible, because the relative phases of ν_1 and ν_2 and of ν_μ and ν_e are arbitrary, rendering the phase angles in (U_{lk}) superfluous.

From the above equations we deduce for the probability of going from one type of neutrino to the other

$$P(l \rightarrow l') = 1/2 \ \sin^2 2\theta \ (1 - \cos\Delta E t) \ . \tag{187a}$$

For kinetic energies much larger than any eigenrest-energies, we can expand

$$E_2 - E_1 \simeq pc\,((1 + m_2^2/2p^2c^2) - (1 + m_1^2/2p^2c^2)) = (m_2^2 - m_1^2)c^4/2pc \tag{188a}$$

to obtain

$$\cos(\Delta E/\hbar)t \simeq \cos 2\pi x/L \tag{188b}$$

$$where \quad L/\hbar c = 4\pi pc/(M_2^2 - m_1^2)c^4 \ . \tag{188c}$$

We see, eq. (187a) may also be written

$$P(l \rightarrow l') = \sin^2 2\theta \, \sin^2 \pi x/L \ . \tag{187b}$$

For convenience, we incorporated ħ and c in the final expressions. Thus 10 MeV neutrinos with a "Δm^2" of 1.6×10^{-10} $(eV)^2$ would have an "oscillation length," L, equal to the radius of Earth's orbit.

From the first announcement by Davis and coworkers that they had detected solar neutrinos (Davis 1968), stating a capture rate of "probably less than three SNU " (Bahcall 1975), that rate has remained remarkably stable. Thirty years later, the published "Homestake" result (from the Homestake mine in Lead, S. D.) was 2.56 ± 0.23 SNU (Cleveland 1998). 1 SNU is a capture rate of one neutrino per second per 10^{36} target atoms. Davis' experiment involves chlorine 37. Two much more recent setups, "SAGE" (Garvin 1997) and "GALLEX" (Anselmann 1995, Hampel 1996), involve Gallium 71 (Kuzmin 1966). Together with "Kamiokande" (from the Kamioka mine near Takayama, Japan) and "Super-Kamiokande" (Fukuda 1996, 1998 (1)), where Cherenkov radiation from knock-on electrons in water is measured, detectors as of the mid 1990s cover three different neutrino energy ranges. The 1998 average event rate at GALLEX and SAGE was 72.3 ± 5.6 SNU, and at Super-Kamiokande the measured boron-8 neutrino flux above 6.5 MeV was 2.44 (1 ± 0.11) x $10^6/cm^2$ sec. Theoretical rates are 129 + 8 and -6, and 5.15 + 1.0 and -0.7, respectively, in the same units as above. For Homestake the theoretical rate is 7.7 + 1.2 and -1.0 SNU (Bahcall 1998 (2)). Clearly, the disagreement between experiment and theory is sizeable.

Theoretical errors are mostly due to errors in nuclear fusion cross-sections, and the treatment of elemental diffusion in the Sun. They are apparently less due to errors in detection cross sections. (Because of the considerably larger error in the Kamiokande than in the Super-Kamiokande data, the former are not included here.)

The use of chlorine 37, first suggested by Pontecorvo (Pontecorvo 1948), involves the reaction

$$\nu_e + Cl^{37} \rightarrow Ar^{37} + e^-$$

the argon decaying back with a 35 day half life. The threshold for the reaction is a relatively high 814 keV. The detection of solar neutrinos by this reaction is therefore limited to "PEP" neutrinos,

$$H^1 + H^1 + e^- \rightarrow H^2 + \nu_e$$

whose energy is 1.44 MeV, to Be^7 neutrinos,

$$Be^7 + e^- \rightarrow Li^7 + \nu_e$$

90% of these with a barely sufficient energy of .86 MeV (the other 10% with .38 MeV, and thus below threshold), neutrinos from C^{13} and N^{15} in the "C-N-O cycle," with endpoint energies of 1.20 and 1.74 MeV, respectively, and, most importantly, the neutrinos from B^8. The decay

$$B^8 \rightarrow Be^8\,(unstable) + e^+ + \nu_e$$

results in neutrinos with an endpoint energy of 14.06 MeV. The by far biggest neutrino flux, that due to the beginning p-p reaction, cannot be utilized. Its maximum energy is .42 MeV. Although in the Sun only about eleven out of ten thousand Beryllium nuclei are converted to B^8, on account of the very high endpoint energy 76% of the calculated response of the Cl^{37} detector is due to B^8 neutrinos (Bahcall 1998 (1)). The transition of Cl^{37} to the analog state in A^{37} (making for a near perfect overlap of wave functions), which has an excitation energy of 4.99 MeV, figures prominently. - It is not the only available excited state. - Davis' detector liquid is perchloroethylene, one hundred thousand gallons of it. The ability to detect some tens of radioactive argon atoms, the result of a few months of operation, belongs in the Guiness book of records. A^{37} has a 35 day half life.

The other "radiochemical experiments" are based on the reaction (Kuzmin 1966)

$$\nu_e + Ga^{71} \rightarrow Ge^{71} + e^-$$

which has the low threshold of 233 keV, and thus also accesses p-p neutrinos. Here the calculated response due to B^8 neutrinos is barely 10% (Bahcall 1998) (1))! The p-p and Be^7 neutrino fluxes can now really make themselves felt. The p-p neutrinos only reach the ground state of Ge^{71}, while two excited states are also within reach of the Be^7 neutrinos; but they do not contribute significantly.

The two gallium detectors are very different. SAGE uses molten gallium, GALLEX uses gallium chloride dissolved in hydrochloric acid.

Measuring the Cerenkov radiation from knock-on electrons, Kamiokande and Super-Kamiokande permit energy and direction measurements, not just plain neutrino detection (Fukuda 1999). Furthermore, they don't just respond to electron neutrinos! But Super Kamiokande only works down to energies of around 6.5 MeV, Kamiokande not even that. Thus, as solar neutrino detectors, they are B^8 neutrino detectors. Theoretically, the neutrino flux from B^8 is the least certain. This is because the cross section of

$$He^3 + He^4 \rightarrow Be^7 + \gamma$$

is only known to within about 10%, and likewise for the cross section of

$$H^1 + Be^7 \rightarrow B^8 + \gamma \ .$$

The B^8 flux is evaluated to be 5.15 x 10^6/cm^2 sec with roughly a 20% error. Contrast this with 5.94 x 10^{10}/cm^2 sec for the p-p generated flux, with a 10% error (Bahcall (1)). - Super Kamiokande contains fifty kilotons of water and over thirteen thousand photo multiplier tubes. It is divided into an inner and outer detector.

Neutrino oscillations provide a mechanism, which may explain the discrepancy between experiment and "straightforward theory". However, the formalism developed in the beginning of this section applies, strictly speaking, only to "vacuum oscillations." Solar neutrinos have to work their way out from the fusion region. But with the terribly small neutrino cross sections, how could this change things much? According to the "MSW mechanism" (Mikheyev 1986, Wolfenstein 1978), it can, greatly. The neutrinos scatter off the electrons, coherently.

As the neutrino beam evolves in a time interval, say dt, it penetrates a layer of matter of thickness dr ≃ cdt. Thus the description of a two-component oscillation will involve the hydrodynamic time derivative of wave functions like those of eqs. (186b), but including the space dependence. to this time derivative will then have to be added the scattering rate.

From eq. (186b) the time rate of change of the electron-neutrino amplitude is found to be given by

$$i\dot{\nu}_e(t) = (E_1\cos^2\theta + E_2\sin^2\theta)\nu_e(t) + \sin\theta\cos\theta\,(E_2 - E_1)\nu_\mu(t)$$

$$= (1/2)[(E_1 + E_2) + (E_1 - E_2)\cos 2\theta\,\nu_e(t) + (E_2 - E_1)\sin 2\theta\,\nu_\mu(t)] \ .$$

Eq. (188a) gives the energy difference in terms of mass-squared difference and neutrino momentum.

The effect on the above expression of replacing $\nu_e(t)$ *by* $\nu_e(x,t)$ and d/dt *by* $\partial/\partial t + (dx/dt)\,\partial/\partial x$ is the replacement of E_1+E_2 by $(m_1^2 + m_2^2)/2p$. Furthermore, $dx/dt \simeq c$. We already know that $E_1 - E_2 \simeq (m_1^2 - m_2^2)/2p$.

To find the scattering rate, we start from a general and important consideration: a plane wave of circular frequency ω is incident normally on an infinite, thin slab of thickness δ. At a distance x from its entry, the wave is, as we know, given by $\exp(ink\delta + ik(x - \delta))$; k is $2\pi/\lambda$, and n is the index of refraction. Attributing the change from the vacuum result to scattering by the slab, we equate (Fermi 1950)

$$\exp(ikx)\exp(i(n-1)k\delta) = \exp(ikx) + N\delta \int_0^\infty (\exp(ikr)/r)\, f(\omega,\Theta) 2\pi\rho d\rho \tag{189a}$$

where $r^2 = \rho^2 + x^2$ and $\tan\Theta = \rho/x$. N is the number of scattering centers per unit volume. (Fermi considered isotropic scattering). To obtain a properly defined integral, we insert, as is - for better or for worse - customary in such cases, the small absorption $\exp(-\varepsilon r)$ in the integrand. Integration by parts yields

$$\int_x^\infty \exp(ikr - \varepsilon r) f(\omega,\Theta) dr = -1/(ik-\varepsilon)\exp(ikx-\varepsilon x) f(\omega,0)$$

$$-1/(ik-\varepsilon) \int_{f(\omega,0)}^{f(\omega,\pi/2)} \exp(ikr-\varepsilon r)\, df(\omega,\Theta)$$

upon which one can let $\varepsilon \to 0$. The second term should go to zero with $1/kx$ (Sakurai 1973). For small $(n-1)k\delta$ we find that

$$\exp(ikx)(1 + i(n-1)k\delta) = \exp(ikx)(1 + i2\pi N f(\omega,0)\,\delta/k)$$

and consequently that

$$n - 1 = 2\pi N f(\omega,0)/k^2 \,. \tag{189b}$$

This is "the optical theorem," apparently first proven by Lorentz.

From $\lim_{n\to\infty} (1 + \xi/n)^n = \exp(\xi)$ it follows that this theorem holds for any slab thickness.

The "forward-scattering amplitude," f(ω,0), has in general a real (scattering) and imaginary (absorbing) part. In the case of neutrinos penetrating the Sun, we are apparently only interested in the real part. The scattering contribution to $i\,\dot{\nu}_e(x(t),t)$ is

$$i\,\dot{\nu}_{sc} = -2\pi N f(\omega,0)/p)\nu_e$$

since we have been using ħ = c = 1. For the charged-current interaction the differential neutrino-electron scattering cross-section in c.o.m. coordinates is (Perkins 1987)

$$(d\sigma/d\Omega)_{c.o.m.} = (G/\pi)^2\, m_e E/2\ .$$

Converting to lab. coordinates (Morrison 1953), and recalling from introductory quantum mechanics that scattering amplitude and differential cross-section are related by

$$d\sigma/d\Omega = |f(\omega,\Theta)|^2$$

one ends up with including the expression $\sqrt{2}\,G\,N\,\nu_e$ in i ν_e (Wolfstein 1995). The transition amplitude for the forward scattering is negative (Langacker 1983). The electron density N is crucial to the neutrino oscillations. Mixing angle and oscillation length are affected by it.

The hydrodynamic time derivative of ν_μ is evaluated similarly to that of ν_e. It differs from that of ν_e by containing a $+\Delta m^2 \cos 2\theta$ and, of course, no charged-current scattering term.

In matrix form we consequently have eq. (190a)

$$i\frac{d}{dt}\begin{pmatrix}\nu_e(x,t)\\ \nu_\mu(x,t)\end{pmatrix} = \frac{1}{4p}\begin{pmatrix} m_1^2+m_2^2-\Delta m^2\cos 2\theta + 4\sqrt{2}NGp & \Delta m^2 \sin 2\theta \\ \Delta m^2\sin 2\theta & m_1^2+m_2^2+\Delta m^2\cos 2\theta\end{pmatrix}\begin{pmatrix}\nu_e(x,t)\\ \nu_\mu(x,t)\end{pmatrix}$$

and for the vacuum eigenstates eq. (190b)

$$i\frac{d}{dt}\begin{pmatrix}\nu_1(x,t)\\ \nu_2(x,t)\end{pmatrix}=\frac{1}{2p}\begin{pmatrix}m_1^2+2\surd 2NGp\cos^2\theta & \surd 2NGp\sin 2\theta\\ \surd 2NGp\sin 2\theta & m_2^2+2\surd 2NGp\sin^2\theta\end{pmatrix}\begin{pmatrix}\nu_1(x,t)\\ \nu_2(x,t)\end{pmatrix}$$

The mixing angle θ is now properly called the "vacuum mixing-angle", θ_v. A similarity transformation of the plain mass-squared matrix of the current neutrinos with the rotation through θ_v results in the diagonal matrix of the mass-eigenstate neutrinos. A similarity transformation of the mass-squared matrix including scattering, see eq. (190a), with the rotation through the "matter mixing-angle" θ_m results in a diagonal matrix for effective neutrino-eigenstates, i.e., states which propagate through matter as plane waves.

From eqs. (188c) we know that the vacuum oscillation-length is (in natural units) $L = 4\pi p/\Delta m^2$. Now there will also be an effective matter-oscillation-length, L_m, made up of the vacuum oscillation-length and Wolfenstein's "characteristic matter-oscillation-length" (Wolfenstein 1978). To emphasize the importance of the density of scattering centers, we shall here denote the latter by L_N,

$$L_N = 2\pi/\surd 2GN \ . \qquad (191a)$$

It is 2π times the inverse of the nonzero eigenvalue of the above matrices (including the respective factor in front) for $m_1 = m_2 = 0$. Recall that the vacuum transition-probability was found to go like

$$\sin^2(\Delta m^2 x/4p) \equiv \sin(\pi x/L_\nu),\ \ (not\ \ 2\pi x/L_\nu).\ \ L_\nu \equiv L\ .$$

The matter mixing-angle may be expressed in the form

$$\tan 2\theta_m = \sin 2\theta_\nu/(\cos 2\theta_\nu - L_\nu/L_N)\ . \qquad (191b)$$

We see that no matter how small θ_v, large θ_m are possible. As a matter of fact, for the condition

$$L_\nu = L_N \cos 2\theta_\nu \tag{191c}$$

there occurs "resonant amplification of neutrino oscillations by matter" (Mikheyev 1986). Without matter, the oscillation length is 2π times the inverse of the difference of the eigenvalues of the mass squared matrix, divided by 2p. This tells us how to get the oscillation length in matter. It turns out to be

$$L_m = L_\nu \sin 2\theta_m / \sin 2\theta_\nu \ .$$

Best least squares fits by Bahcall, Krastev and Smirnov (Bahcall 1998 (2)) to the Homestake, the weighted average of GALLEX and SAGE, and the Super-Kamiokande data resulted in the small θ and a less probable large θ solution:

$$\sin^2 2\theta = 6.0 \, x \, 10^{-3} \, , \; \Delta m^2 = 5.4 \, x \, 10^{-6} (eV)^2$$
$$and \; \sin^2 2\theta = 0.76 \, , \; \Delta m^2 = 1.8 \, x \, 10^{-5} (eV)^2 \ .$$

Our use of electron and muon-neutrinos in the discussion of vacuum and matter oscillations should not be construed to imply that these are the dominant modes in solar neutrino propagation.

On to reactor experiments. The U^{235} equilibrium-fission neutrino spectrum peaks a little below 1 MeV. By 3.5 MeV it is down to one tenth of its maximum value, and at 6 MeV to less than one hundredth.

Oscillation experiments near reactors may be set up at various distances from the core. The most frequently used detection reaction,

$$\nu_a + p \rightarrow e^+ + n$$

has a cross section of roughly 10^{-43} cm^2 per antineutrino, the actual value depending on the fuel element composition at the time of measurement (Boehm 1987). An experiment known as the "Bugey experiment" (Cavaignac 1984) has been interpreted positively. The corrected yield at 18.3 meters was 10% lower than expected, when compared to the yield at 13.6 meters. Together with spectral measurements, this lead to the preferred parameters $\sin^2\theta = 0.25$ and $\Delta m^2 = 0.2$ $(eV)^2$. Other experiments gave negative results (Boehm 1987).

Using the antineutrino disintegration of the deuteron, Reines determined the "ratio of ratios," $(\sigma_{cc}/\sigma_{nc})_{obs}/(\sigma_{cc}/\sigma_{nc})_{calc} = R$, in his search for oscillations. Here cc and nc refer to the reactions

$$\nu_a + d \to 2n + e^+ \; and \; \nu_a + d \to n + p + \nu_a \; .$$

Using heavy water as target, the value obtained for R was R = 0.74 ± 0.23, within one standard deviation (Reines 1983). The charged-current interaction products were obviously not noticeably depleted. What about the neutral-current interaction products?

The deuteron disintegration is perhaps the first investigated process bearing on the coherent scattering of neutrinos (Ahrens 1970). In the charged mode (charge exchange process) the incoming antineutrino interacts with the proton states of nucleon one and nucleon two, leading to positron emission

$$(\sigma_1\tau_1^- + \sigma_2\tau_2^-)(1(p)2(n) - 1(n)2(p))/\sqrt{2}$$
$$= (\sigma_1 - \sigma_2)\,1(n)2(n)/\sqrt{2}$$

while in the neutral mode (charge retention) it interacts with the proton and neutron states of each nucleon, leading to the reemission of an antineutrino

$$(\sigma_1\tau_1^3 + \sigma_2\tau_2^3)(1(p)2(n) - 1(n)2(p)/\sqrt{2}$$
$$= (\sigma_1 - \sigma_2)(1(p)2(n) + 1(n)2(p))/\sqrt{2} \; .$$

The final state wave functions in isospace are

$$1(n)2(n) \;\; and \;\; (1(p)2(n) + 1(n)2(p))/\sqrt{2}$$

respectively. The $\boldsymbol{\sigma}_1$ and $\boldsymbol{\sigma}_2$ produce allowed (no orbital angular momentum being carried off) Gamov-Teller transitions. Note, that the subscripts refer to the number assigned to each nucleon. Also, the effective neutral-current interaction is one of isovectors. We conclude that coherence between scattering off particles one and two lets the isospin contribution to the two transition probabilities differ by a factor of two, in favor of the neutral mode. Of course, this basic fact is overcompensated by another basic fact: the square of the effective coupling constant of the charged mode is four times the one of the neutral mode.

In case the reader is not familiar with the term "Gamow-Teller transition": nuclear β-decay is due to vector and axial vector interactions, γ^μ and i $\gamma^\mu\gamma_5$. In distinction from purely leptonic (second order) processes, processes involving nuclei have an axial vector coupling constant different from G. Thus for the decay of the neutron it is (Perkins 1987)

$$G_A = 1.25\, G \ .$$

While the vector current is conserved (Gerstein 1955, 1957, Feynman 1958, Sudarshan 1957, Sakurai 1958), the axial-vector current is not. Whenever a current engaging in a certain interaction is conserved, other interactions do not affect the coupling constant concerned (Okubo 1959). The axial-vector coupling constant is affected by strong interactions, and the above constant is surprisingly well accounted for (Adler 1965, Weisberger 1966).

Gamow-Teller transitions derive from the axial-vector interaction: together with the iβ from the $\bar{\psi}$, the i $\gamma^k\gamma_5$ gives an i σ^k, in both the nuclear and lepton covariants of the entire (second order but low energy) matrix element. This explains the appearance above of **σ** in the nuclear part of the deuteron disintegration, one for each nucleon.

With a fission spectrum, the cross sections are of order

$$\sigma_{cc} \approx 2\, x\, 10^{-45}\, cm^2 \;\; and \;\; \sigma_{nc} \approx 5\, x\, 10^{-45}\, cm^2 \ .$$

Theory and experiment are in good agreement on this. As to fluxes, the Reines experiments, e.g., were performed near the Savannah river reactor in a flux of about 2.5 x 10^{13} antineutrino/cm^2 s. - Why is σ_{cc} smaller than σ_{nc}?

The use of the ratios of "cc" and "nc" events is quite clever: theoretical uncertainties of neutrino flux and spectrum tend to cancel in the calculated ratio, and since the total rate of neutral-current interactions is not affected by oscillations, the experimental ratio is a normalized measure of oscillations. But in these deuteron disintegration experiments only the neutrons were detected, using He^3 proportional counters immersed in the heavy-water container. This required estimating what proportion of one-neutron events were actually due to the cc interaction. The error in R is correspondingly large. - Cross sections differential in proton or neutron energy have been calculated (Ahrens 1979), and may prove useful in more differentiated experiments based on the deuteron disintegration (Lang 1978). -

Reactor experiments based on the antineutrino on proton into neutron and positron reaction have been very differentiated. The basic scheme followed at several sites, including Bugey, involves the recording of time correlated neutron and positron events in separate detectors. For neutrons He^3 contained in wire chambers, permitting localization of captures; liquid scintillators for positron detection and energy determination. Pulse-shape discrimination distinguishes positrons from knock-on protons (Kwon 1981).

While the two reactor experiments described above are mostly of historical interest, a present day reactor experiment of great interest is the CHOOZ experiment in the Ardennes. The setup is 1 Km distant from two reactors. The detection is based on antineutrinos on protons, the medium being a gadolinium loaded liquid scintillator. The resulting positrons and neutrons are observed in delayed coincidence, the neutrons by the 8 MeV γ-emission following their capture in gadolinium. Detection of the positrons in the scintillator is supported by detection of their annihilation gammas. Here we merely quote two limits:

$$(\Delta m)^2 > 0.9\,x\,10^{-3}\,(eV)^2$$

is excluded for maximum mixing,

$$\sin^2 2\theta > 0.18$$

is excluded for $(\Delta m)^2 \geq 2 \times 10^{-2}$ $(eV)^2$ (Apollonio 1998, Bemporad 1999).

During the nineteen-nineties atmospheric neutrino studies have played a role of rapidly increasing importance. The ν_μ/ν_e flux ratio has been calculated by various groups and (without oscillations) is thought to be known to better than 5% for energies between 100 MeV and 10 GeV (Agraval 1996, Gaisser 1996). The agreed on calculated value is about two for energies ≤ 1 GeV, and thereafter increases with neutrino energy. The value of two is essentially due to the predominant neutrino production mechanisms

$$\pi^{\pm} \to \mu^{\pm} = {\nu_{\mu} \atop \bar{\nu}_{\mu}} \quad and \quad \mu^{\pm} \to e^{\pm} + {\bar{\nu}_{\mu} \atop \nu_{\mu}} + {\nu_{e} \atop \bar{\nu}_{e}}$$

the bars denoting antineutrinos. Kaons contribute less than 10% to neutrinos responsible for "contained events" (in the inner detector), i.e., almost all electron neutrino and the lower energy muon neutrino caused events (Gaisser 1996).

In Kamiokande and Super Kamiokande experiments it has become customary to also use a ratio of ratios as the principal reported quantiry,

$$R = (\mu/e)_{DATA}/(\mu/e)_{MC}$$

MC meaning Monte Carlo simulation; μ and e are the number of μ-like and e-like events, respectively. These result mostly from charged-current interactions of the neutrinos with nuclei (Fukuda 1998(2) a. (3)). Neutral-current interactions with nuclei can produce charged leptons via pion production, in which case the recorded charged lepton does not imply, that the responsible incoming neutrino was of the same generation. Charged-current interactions can do the same, but with much less likelihood than the direct production of charged leptons. A much greater percentage of e-like events is due to these complex processes than μ-like events (Fukuda 1998(3)).

In each ratio uncertainties tend to cancel, experimental ones and total flux and spectral uncertainties, respectively. For energies sufficiently above the muon mass, so do errors in neutrino cross sections. The uncertainty in flux is particularly noteworthy. It is mostly due to the different treatments of pion production. Neutrinos in the GeV range are largely due to pions produced by 10 to 30 GeV protons (Gaisser 1996). The primary proton spectrum itself is uncertain by about 10%. The interpretation of anomalies in R may well be dependent on one's knowledge (or assumptions) of absolute fluxes and spectra. As a matter of fact, uncertainty in calculated spectra has once lead to the interesting suggestion that some events could be due to positrons from the decay of protons (Mann 1992).

Already after the first year and a half of operation, the atmospheric neutrino data from Super-Kamiokande proved decisive. Values of R of 0.63 and 0.65 to an accuracy of about 10% have been reported for sub-GeV and multi-GeV events, respectively (Fukuda 1998 (2)). These values have rather generally been accepted to imply massive (as opposed to massless) neutrinos, via the oscillation mechanism. Even more striking, there is a conspicuous dependence of the μ-like data on zenith angle. The asymmetry parameter

$$A = (U - D)/(U + D)$$

with U the number of "upward going events", events with a cosine of zenith angle between -1 and -0.2, and D the number of "downward" ones, events with a cosine between 0.2 and 1, has been plotted by Fukuda et al. as a function of momentum. The most outstanding result: while e-like events are consistent with no asymmetry, multi-GeV μ-like events, defined to be events due to μ neutrinos with a momentum greater than about 1400 MeV/c, are definitely asymmetric. (Beyond 1 GeV/c, geomagnetic effects on cosmic rays are small.) Averaging over contained and partially contained multi-GeV μ-like events, the data give

$$A = -0.296$$

with an error of about 15%. Thus U is approximately 43% of D. Viewed as engaging in a two-component oscillation, the muon neutrinos, upon traversing the Earth, have established an equilibrium with tau neutrinos, or perhaps sterile neutrinos (see end of Section 6), with a mixing angle of roughly 45 degrees. Fukuda et al. get a best fit with the parameters $\sin^2 2\theta = 1$ and $\Delta m^2 = 2.2 \times 10^{-3}$ $(eV)^2$. The measured electron-neutrino flux is roughly 20% higher than that found from Monte Carlo calculations. Interestingly, the various calculated neutrino fluxes vary by about ± 20 %. - A three-neutrino-model calculation gives a best fit close to the one above. It allows some $\nu_\mu \rightleftarrows \nu_e$ and $\nu_e \rightleftarrows \nu_\tau$ oscillations (Barger 1998).

Of the neutrino-oscillation experiments performed at accelerators, we mention the much discussed $\bar{\nu}_\mu \rightleftarrows \bar{\nu}_e$ *and* $\nu_\mu \rightleftarrows \nu_e$ measurements at the Los Alamos Meson Physics Facility, an 800 MeV Linac. They resulted in oscillation probabilities in percent of 0.34 plus minus 0.12 and 0.26 plus minus 0.10, respectively. Strictly speaking, this assumes that the $\bar{\nu}_e$ and ν_e excesses above estimated backgrounds are due to oscillations, and not to direct violation of "lepton-number conservation" (Athanassopoulos 1996, 1998 and Hill 1995). By this violation one means the possible admixture of the current $\bar{\psi}_e \gamma^\mu (1+\gamma_5)\psi_{\bar{\nu}}$, see sections 3.5 and 7.3, which would lead to electron neutrinos and, by its hermitian conjugate, to electron antineutrinos. Δm^2 fits to the data cover a wide range. For maximum mixing .02 eV$\leq \Delta m^2 \leq$ 0.1 eV, from $\bar{\nu}_\mu \to \bar{\nu}_e$, 90%C.L.

The muon-antineutrinos come from the decay at rest $\mu^+ \to e^+ + \nu_e + \bar{\nu}_\mu$, the muons themselves coming from $\pi^+ \to \mu^+ + \nu_\mu$. The pi mesons are produced in a water target by the 800 MeV protons of the linac, while the decays take place in a copper beam stop. The electron antineutrinos, if any, are detected by the "LSND" (the liquid scintillation neutrino detector) through the usual $\bar{\nu}_e + p \to e^+ + n$, followed by the capture gammas of $n + p \to d + \gamma$. The LSND is located 30 m behind the beam stop. The muon neutrinos result from π^+ decay in flight. The electron neutrinos, if any, are then detected by the LSND through the production of electrons in the mineral oil. Because the muon neutrinos come from the π^+ decay, lepton-number nonconservation cannot be responsible for the appearance of electron neutrinos!

Let us briefly look at the generalization from two to three kinds of neutrino flavors. The special unitary matrix of eq. (184) is a three-by-three matrix belonging to the group SU(3). As SU(2) has three parameters, so SU(3) has eight. (SU(n) has n^2-1 parameters) [A.11]. Remember, with two states transforming, the relative phase between those states and that between the transformed states are arbitrary, leaving only one angle to be determined. With three states transforming, two relative phases between them and two between the transformed states are arbitrary. This leaves four of the original eight parameters, one more than the three angles of a three-by-three rotation matrix (Kane 1987).

7.2.1 Different Three-Component Bases of Quarks and of Neutrinos

In 1973 Kobayashi and Maskawa proposed just such a matrix for transforming hadronic quarks into quarks which are eigenstates of the weak interaction. By "hadrons" are meant baryons and mesons. They introduced the three-by-three "mixing matrix" even though charmed quarks, proposed theoretically (Glashow 1970), had not yet been found experimentally, not to mention bottom and top quarks. In this way a reason was also given for the violation of CP invariance in the decay of neutral K-mesons (Christenson 1964). This is so because the fourth parameter in the mixing matrix which connects down, strange and bottom quarks of the two bases (see eqs. 192), is a phase angle, causing the transition amplitudes for the decaying mesons to be complex.

Already during our first discussion of the Dirac equation we learned, that time reversal involves complex conjugation, and so the just mentioned transition amplitude is not invariant under time reversal! There is a famous theorem, "the CPT theorem", which is a statement of the conditions under which a field theory is invariant under the combined operations of charge conjugation, space inversion and time reversal (Lüders 1957, Pauli 1957). Practically all present day field theories satisfy those conditions. Therefore a violation of T invariance implies a violation of CP invariance. In 1988 it was established to 3.5 standard deviations, that the standard model of electroweak theory and chromodynamics, the latter the theory of quarks and "gluons", includes an explanation of the observed violation of CP invariance (Burkhardt 1988, Woods 1988).

The three-by-three matrix connecting neutrino mass-eigenstates to weak-interaction eigenstates, the two sets presumed to be different from one another, is a lepton analogue of the "Cabibbo-Kobayashi-Maskawa matrix" (Rev. of Particle Phys. 1998). But while this matrix appears in the calculation of decay properties, the neutrino analogue does not, since there are no charged particles into which the three flavor-neutrinos can decay. As regards oscillations, there is the well known case of the neutral K-mesons, where strongly produced K° mesons make weak transitions to their antiparticles (see also section 5.6),

$$K^o \rightleftharpoons \bar{K}^o .$$

There are weak-interaction eigenstates (Gell-Mann 1955, Pais 1955, Sachs 1956).

$$K_1^0 = (K^0 + \bar{K}^0)/\surd 2 \quad and \quad K_2^0 = (K^0 - \bar{K}^0)/\surd 2$$

disregarding the small correction due to CP nonconservation. With their small half widths, these states are also close to being mass eigenstates. Keep in mind: here we are dealing with two-particle systems (quark-antiquark systems). The weak-interaction eigenstates are not stable, as they are (except for oscillations) in the neutrino case. They decay exponentially. The detection of the products of oscillation and decay proceeds via the strong or electromagnetic interactions, while any neutrino-oscillation products can only be detected weakly. - This superposition of different particles, at the time considered to be elementary, is historically the first example of experimentally verifiable transformations from one set of particle states to another. (Including CP nonconservation, see Lee 1957).

Commonly the SU(3) transformations are applied to sets of quarks or leptons whose members have the same isospin component. The fact that this component is down for quarks and up for leptons - neutrinos have been chosen to be up - is purely conventional. Since the transformations are elements of a group, a nondiscriminating application of two transformations to a current, one to the initial and one to the final state, is equivalent to a transformation of just one of the states. Take the case of the charged-quark currents

$$j_w^\mu = g/2\,(u,c,t)_L\, U'^+ \gamma^\mu U'' \begin{pmatrix} d \\ s \\ b \end{pmatrix}_L \tag{192a}$$

where

$$U' \begin{pmatrix} u \\ c \\ t \end{pmatrix}_L = \begin{pmatrix} u \\ c \\ t \end{pmatrix}_{WL} \quad and \quad U'' \begin{pmatrix} d \\ s \\ b \end{pmatrix}_L = \begin{pmatrix} d \\ s \\ b \end{pmatrix}_{WL} \tag{192b}$$

are weak-eigenstate quarks. Clearly

$$U'^+ U'' = U'^{-1} U'' = U \ . \tag{193}$$

As before, L stands for left handed. The primes are supposed to point out that the matrices, to which they are affixed, depend on different parameter values. Indeed, when he first related the mean lives of strangeness changing and strangeness nonchanging semi-leptonic decays, e.g., of $\Sigma^- \to n + e^- + \bar{\nu}_e$ and $n \to p + e^- + \bar{\nu}_e$ and of $K^- \to \pi^0 + e^- + \bar{\nu}_e$ and $\pi^- \to \pi^0 + e^- + \bar{\nu}_e$, Cabibbo considered the hadron currents to be transformed as a whole. The coupling constants of these decays were related to the coupling constant of the decay of the muon by the multiplication with $\sin\theta$ and $\cos\theta$, respectively (Cabibbo 1963). θ_C is known as the "Cabibbo angle". Correcting for the "nonconservation of the axial-vector current" (Taylor 1958, Goldberger 1958, Weinberg 1998) and for quark masses, it is in terms of its cosine (Review of Particle Physics 1996)

$$0.9745 < \cos\theta_C < 0.9757 \ . \tag{194}$$

The decays mentioned above contain no examples with neutral currents. The reason: strangeness changing neutral currents (charge retention currents) do not exist. For at least the entire decade of the sixties this was viewed by many as militating against neutral currents altogether. The introduction of a fourth quark into weak-interaction theory provided a reason for the nonobservation of neutral currents of the strangeness changing variety, without predicting such for the strangeness conserving kind (Glashow 1970). This is seen immediately, if in eqs. (192) the weak I-spin-up quarks are replaced by their down partners. With down quarks in final and initial states, the product $U'^{\dagger} U''$ equals one. Weak-eigenstate and hadronic quarks give the same neutral currents.

7.3. PARTICLE-PHYSICS ASPECTS OF DOUBLE-BETA DECAY

Goeppert-Mayer first calculated the probability of double-beta decay, and, for a mass difference between parent and daughter of .002 mass units, found a lifetime of roughly 10^{25} years (Goeppert-Mayer 1935). This result provided a good explanation for the large number of isobaric pairs with an atomic number difference of two. Then, not long after the publication of Majorana's article on symmetry properties of the Dirac equation (Majorana 1937), Furry described a variant of the process made possible by the self-charge-conjugate neutrino proposed in Majorana's article: "neutrinoless double-beta decay," abbreviated as 0ν $\beta\beta$-decay, see Fig. 3. Furry calculated lifetimes shorter by a factor of 10^5 to 10^{15} years (Furry 1939). The latter lifetimes allowed the conclusion that 0ν $\beta\beta$-decay may have a bearing on isotope abundances and long lived radioactivities. (For chronological listings of geochemical half life determinations and counter experiments, see Moe 1994).

As already alluded to in section 3.5., 0ν $\beta\beta$-decay requires Majorana neutrinos. In the absence of a right-handed lepton current, e.g. (1/2) $\bar{\psi}_e \gamma^\mu (1 + \gamma_5) \psi_{\bar{\nu}}$, 0ν $\beta\beta$-decay obviously requires the Majorana neutrino to be massive. Only then will one handedness contain both helicities, see eqs. (23), and thus the emitted neutrino have a chance of being reabsorbed in the double-decay process: the projection of the wave function

$$-i\bar{\psi}_{\nu R} = ((1-\gamma_5)\beta\psi_\nu)^\dagger/2 \qquad (195a)$$

on

$$\psi_{\nu L} = (1-\gamma_5)/2\,\psi_\nu \qquad (195b)$$

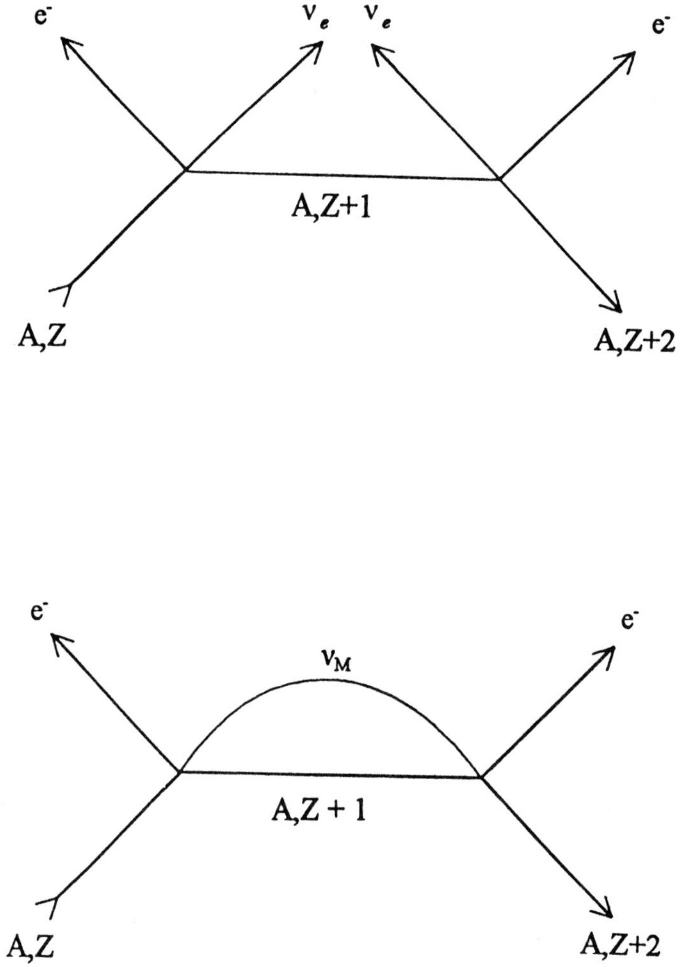

Figure 3: Feynman diagrams for 2ν ββ-decay and 0ν ββ-decay. The vector-boson lines have been omitted.

equals $m/2p^0$, see eqs. (12a) and (12d), leading for non vanishing mass to the reabsorption of emitted right-handed neutrinos. Should there be a right-handed lepton current like the one mentioned above, a mass may well be required for gauge theories (Schechter 1981, Kayser 1987).

The spin-one-half mass term in the Lagrangian density may be written

$$i\mathcal{L}_m = -m(\bar{\psi}_R\psi_L + \bar{\psi}_L\psi_R),\ \psi = \psi(x^\mu)$$

check end of section 6. We know that weak interactions involve left-handed currents, so we modify this mass term in the case of neutrinos by attaching a prime to the right handed functions. These may have some coupling too. Thus quite generally

$$i\mathcal{L}_m = -(m\,\bar{\psi}_R{}'\psi_L - h.c.)\ . \tag{196}$$

This $\mathcal{L}_m$ shall be reformulated with the help of the following relations:

$$\begin{aligned}\psi_R^c \equiv (\psi_L)^c &= ((1-\gamma_5)/2\ \psi)^c \\ &= (1+\gamma_5)/2\ \psi^c\end{aligned} \tag{197a}$$

$$\begin{aligned}\bar{\psi}_R^c &= ((1-\gamma_5)/2\ \psi)^{c\dagger}\gamma^{0\dagger} \\ &= \bar{\psi}^c(1-\gamma_5)/2\end{aligned} \tag{197b}$$

and similarly for ψ_L^c and $\bar{\psi}_L^c$.

Then

$$\begin{aligned}\bar{\psi}_R^c\,\psi_L^{c\,\prime} &= \psi^T\gamma_2\gamma^{0\dagger}(1-\gamma_5)/2\ \gamma_2\ \psi^{*\prime} \\ &= -\ \psi^T\gamma^{0\dagger}(1+\gamma_5)/2\ \psi^{*\prime} \\ &= -\ \psi^T(1-\gamma_5)/2\ \gamma^{0\dagger}\ \psi^{*\prime}\ .\end{aligned}$$

Transposition results in

$$\overline{\psi}_R^c \, \psi_L^{c\,\prime} = \overline{\psi}_R{}' \, \psi_L \; . \tag{198a}$$

In making a change of sign in the last step we borrowed from quantum field theory. We treated the ψ functions as operators, in which case they anticommute, since they describe spin-one-half fields. Systems of spin-one-half particles obey the Pauli principle. As we know from nonrelativistic quantum mechanics, their wavefunctions must be antisymmetric under an exchange of the space and spin coordinates of two particles. This antisymmetry is reflected in the sign change, when the two operators are exchanged.

Similarly to the derivation of equation (198a) we derive

$$\overline{\psi}_L^{c\,\prime} \, \psi_R^c = \overline{\psi}_L \, \psi_R{}' \; . \tag{198b}$$

We may now write

$$i\,\mathcal{L}_m = -1/2\,(m\overline{\psi}_R{}'\,\psi_L + m^*\overline{\psi}_L^{c\,\prime}\,\psi_R^c - h.c.) \; . \tag{199a}$$

But what about the Majorana neutrino? There the left-handed components ψ_L and right-handed components ψ_R^c, and the right-handed components $\psi_R{}'$ and left-handed components $\psi_L^{c\,\prime}$ form the 4-spinors. Thus the pertinent Lagrangian density is

$$i\,\mathcal{L}_M = -1/2\,(M_L\overline{\psi}_R^c\psi_L + M_R\overline{\psi}_L^{c\,\prime}\psi'_R - h.c.) \; . \tag{199b}$$

There are two masses. This is not surprising. Two Majorana-neutrino fields (in the sense of the function of eq. (64)) have the same number of degrees of freedom as one Dirac-neutrino field.

Now, for a Majorana neutrino one should really consider two cases,

$$\psi_M(x^\mu) = \pm\psi_M^c(x^\mu) \tag{200}$$

although in our discussion we had no reason up to now to consider but one case, the equality of ψ_M and ψ_M^c . However, with the possibility of two neutrinos contributing to the same reaction, neutrinos which just may be Majorana neutrinos, the sign becomes essential. One will have to distinguish between Majorana and, to use an archaic term, anti-Majorana neutrinos.

Adding the two Lagrangian densities shows there to be a "mass matrix"

$$\mathbf{M} = \begin{pmatrix} 0 & 0 & M_L & m \\ 0 & 0 & m & M_R^* \\ M_L^* & m^* & 0 & 0 \\ m^* & M_R & 0 & 0 \end{pmatrix} . \qquad (201a)$$

The column vector which this matrix multiplies is

$$\psi = \begin{pmatrix} \psi_R^c \\ \psi_R{}' \\ \psi_L \\ \psi_L^{c\,\prime} \end{pmatrix} . \qquad (201b)$$

The row vector is $\overline{\psi}$. Massive Majorana neutrinos satisfy the Dirac equation. With diagonal γ_5 , we may write a general Majorana neutrino field

$$\psi_M = \begin{pmatrix} \pm\,\psi_R^c \\ \psi_L \end{pmatrix} , \quad \psi_R^c \equiv (\psi_L)^c \qquad (202)$$

the relative minus sign applying to anti-Majorana neutrinos. Consequently, masses in the matrix can be negative. Furthermore, if $\mathcal{L}_m$ is not CP invariant, the two symmetric submatrices are complex. - Primakoff considered the possibility of the complex coupling of a right-handed lepton current involving Majorana neutrinos, thus violating CP and lepton-number conservation at the same time

(Primakoff 1969). (A generalization to where right-handed nucleon currents are included can be found in Haxton 1984).

Rather than proceed piecemeal, we right away generalize to allow inclusion of two or more flavors. Thus terms like $\bar{\psi}_R{}'\, m\, \psi_L$ and $\bar{\psi}^c_R M_L \psi_L$ will in general imply double sums in flavor space, like $\bar{\psi}'_{Rk} m_{kl} \psi_{Ll}$ and $\bar{\psi}^c_{Rk} M_{kl} \psi_{Ll}$, respectively. The mass matrix will then read

$$\mathbf{M} = \begin{pmatrix} 0 & 0 & M_L & m^T \\ 0 & 0 & m & M_R^\dagger \\ M_L^\dagger & m^\dagger & 0 & 0 \\ m^* & M_R & 0 & 0 \end{pmatrix} . \tag{203}$$

Clearly, **M** is hermitian. We shall follow the route taken by Haxton and Stephenson (Haxton 1984), and diagonalize the two submatrices, which we shall call M and $M^\dagger$, first. From eqs. (199) we gather that in the Lagrangian density M is sandwiched between $-i(\bar{\psi}^c_R, \bar{\psi}_R{}') \equiv -i\bar{\Psi}^c_R$ and $\begin{pmatrix} \psi_L \\ \psi_L^{c\prime} \end{pmatrix} \equiv \Psi_L$.

Thus

$$\bar{\Psi}^c_R M \Psi_L = \overline{U^T \Psi^c_R} D\, U^{-1} \Psi_L = \bar{\Psi}^c_R U^* D U^{-1} \Psi_L = \bar{\nu}_R D\, \nu_L \tag{204a}$$

where D is the diagonalized M. U, the eigenvector matrix, is taken to be unitary. As far as their transformation properties under U are concerned, Ψ^c_R and Ψ_L differ only by complex conjugation. We defined $\nu_L \equiv (\nu_R)^c$. - Do not confuse these ν_L and ν_R with the current or flavor ones. - For the submatrix $M^\dagger$

$$\bar{\Psi}_L M^\dagger \Psi^c_R = \bar{\Psi}_L U D\, U^T \Psi^c_R = \bar{\nu}_L D\, \nu_R \tag{204b}$$

follows. The second step is then to diagonalize

$$\begin{pmatrix} 0 & D \\ D & 0 \end{pmatrix} \rightarrow \begin{pmatrix} D & 0 \\ 0 & -D \end{pmatrix} = \mathbf{D} . \tag{205}$$

The matrix **D** is obviously not positive definite. I.e., for every mass eigenvalue m_k, positive or negative, there is a $-m_k$. The "eigenfields" are the Majorana and anti-Majorana neutrino functions

$$\nu^+ = \nu_L + \nu_R \ and \ \nu^- = \nu_L - \nu_R \tag{206}$$

as one can see from the right most terms in eqs. (204).

For only one flavor there are two ν^+ and two ν^- eigenfields, ν_k^+ and ν_k^-, k = 1,2. This is clear from the mentioned last terms in eqs. (204). E. g., $\bar{\nu}_{Rl}$ and ν_{Ll} multiply m_l and so do $\bar{\nu}_{Ll}$ and ν_{Rl} . For all three presently known flavors there consequently are six ν^+ and six ν^- fields. The subscript k goes from one to six. In the diagonalized twelve-by-twelve mass matrix, the top six eigenvalues go with the ν_k^+ , the bottom six with the ν_k^- . The current or flavor fields are obtainable from the fields ν_L and ν_R by

$$\Psi_L = U \nu_L \quad and \quad \Psi_R^c = U^* \nu_R \tag{207}$$

see eqs. (204). We called the right-handed eigenfields ν_R rather than ν_R^c , so as to adhere to the convention (Haxton 1984). The left-handed current neutrinos (as opposed to antineutrinos) are then given by

$$\psi_{Lk} = U_{kl} \nu_{Ll} \quad , \quad k = 1 \ to \ 3 \ , \ l = 1 \ to \ 6 \tag{208a}$$

and the right handed by

$$\psi'_{Rk} = U^*_{k'l} \nu_{Rl} \ , \ k = 1 \ to \ 3 \ , k' = k+3 \ , l = 1 \ to \ 6 \ . \tag{208b}$$

Recall that the fields ν_L and ν_R are superpositions of the lefthanded ψ_L and $\psi_L^{c\prime}$

and of the righthanded ψ_R^c and $\psi_R{}'$ respectively. With three flavors, each of these last four fields has three components (in flavor space).

Just as in ordinary β decay theory, one factors the transition probability of a ββ decay into the coupling constant squared (now G^4), the Coulomb correction, a matrix element squared, and the phase-space-factor. The first three factors come from the square of the transition amplitude. ββ decay, treated by perturbation theory, has as part of its matrix element the usual denominator of intermediate minus initial energy, to be summed, together with the numerator, over all intermediate states. This is where 0ν ββ-decay is at the same time easier and harder to treat than 2ν ββ-decay. Easier, because the average virtual neutrino energy is considerably higher, 50 or so MeV, than the average nuclear excitation energy, say 10 MeV, let alone the energy of a β particle.

Thus, so as to be able to sum over a complete set of unweighted nuclear states (the nuclear part of the intermediate states), the rather standard approximation of using an average nuclear excitation energy can be made with reasonable confidence. Harder, because of the factor in the matrix element accounting for the emission and absorption of the virtual neutrino by pairs of neutrons. Considering this neutrino to be a superposition of mass eigenstates means that the energy dependence of the denominator will include a dependence on the eigenmasses. However, for "light neutrinos", i.e., for neutrinos of mass small compared to the range of intermediate neutrino energies, the mentioned factor is usually integrated over the neutrino momentum after setting the eigenmass in the denominator equal to zero. As a consequence the eigenmasses will only appear outside the matrix element. Now, each eigenmass carries a weight determined by the fraction it contributes to the mass of the current neutrino concerned, in this case the electron antineutrino. We express this as

$$U_{1k} m_k U_{kl}^T \, |ME(m_k)| \rightarrow U_{1k} m_k U_{1k} \, |ME| \; . \tag{209}$$

The transformed eigenmasses in this equation give the matrix element $M_{11}^\dagger$, see eq. (203). We see this if we remember that

$$D = U^T M U$$

and that the Lagrangian mass density of a righthanded antineutrino of the Majorana variety (identifiable by the "destroyed components") is $-i/2\, \bar{\psi}_{Lk} M_{Lkl}^\dagger \psi_{Rl}^c$, no sums. Belonging to the first generation, the left-handed electron neutrino and right-handed antineutrino get the subscript one. Thus

$$m_\nu = m_k U_{1k} U_{1k} \cdot \tag{210a}$$

Remember the approximations made!

The eigenmasses of one submatrix need not all be positive. But any mass m_k can be converted into its negative by "mass conjugation", i.e., by taking a γ_5 through the Dirac operator,

$$\gamma_5(\alpha\cdot\mathbf{p}_{op} + \beta m)\psi = (\alpha\cdot\mathbf{p}_{op} - \beta m)\gamma_5\psi = 0 \ . \tag{211}$$

If ψ satisfies the Dirac equation, then a ψ' with the opposite relative sign between left and right-handed components satisfies the mass conjugate equation. For Majorana (anti-Majorana) neutrinos we then have

$$\begin{aligned} \gamma_5\psi_M(m) &= \gamma_5(\psi(m) \pm \psi^c(m))/2, \\ &= (\psi(-m) \mp \gamma_2\psi^*(-m))/2 \ \mathit{because}\ \gamma_5\gamma_2 = -\gamma_2\gamma_5 \\ &= (\psi(-m) \mp \psi^c(-m))/2 \ . \end{aligned} \tag{212}$$

The signs of the phase factor $\eta^c = \pm 1$, defined by

$$C\psi_M = \eta^c\psi_M \tag{213}$$

are inverted by γ_5 conjugation of ψ_M.

We know that for every m_k associated with ν^+ there is a $-m_k$ associated with ν^-. Thus choosing only positive eigenmasses, eq. (210a) can be rewritten

$$m_\nu = \eta_k^c m_k U_{1k} U_{1k} \ . \tag{210b}$$

Any positive m_k coming from the lower half of the matrix **D** multiplies the charge-conjugation phase-factor $\eta^c = -1$. Wolfenstein observed the significance of the phase factor produced by the operation CP (Wolfenstein 1981). Under space inversion one has

$$\psi(x) \to \psi'(x) = \eta^P \beta\, \psi(-x)$$

see eq. (30d). Consequently, using a Dirac ψ,

$$\begin{aligned}\psi(x) \pm \psi^c(x) &\to \eta^P \beta\, \psi(-x) \pm \eta^{P*} \gamma_2 \beta \psi^*(-x) \\ &= \beta(\eta^P \psi(-x) \mp \eta^{P*} \psi^c(-x)) \ .\end{aligned} \tag{214}$$

Thus the choice $\eta^P = i$ (or $-i$) makes the relative sign of the phase factors of Majorana and anti-Majorana neutrino under CP depend only on C: with a CP invariant weak interaction the mass eigenfields are also eigenfields of the weak interaction (Remember the neutral K-mesons).

- An imaginary phase factor was used in the early work on the universal Fermi interaction (Yang 1950). -

Experimental work in double-beta decay has been going on since 1948. Direct methods, meaning detection of the decay in process and, possibly, actual measurement of the electron energies, were not crowned with success for nearly forty years. The counting measurement of the half life of selenium 82 by Moe and collaborators, (1.1 + 0.8 and -0.3) x 10^{20} years , was the first direct observation of 2ν ββ-decay (Elliott 1987). Before that, the existence of 2ν ββ-decay had been rather unambiguously established by geochemical means. These involve analyzing minerals of presumably known age, containing an isotope (A,Z), for the presence of the daughter (A,Z+2). The half life of tellurium 130 had thus been established to be $T_{1/2} \approx 3 \times 10^{21}$ y (Hayden 1953) and 8×10^{20} y (Takaoka 1966).

The Moe experiment involved a time projection chamber (Moe 1984). Observation of the drifting ionization electrons in the chamber gas, produced by the decay electrons, allows measurement of the latter's energy, not just of their occurrence.

A great advance was that of Fiorini and coworkers, who introduced source and detector as one in form of a germanium crystal (Fiorini 1967). Natural germanium contains 7.78% Ge^{76}, which decays into selenium 76; $\Delta E \sim 2.05$ MeV. Ge(Li) detectors have an energy resolution of a few KeV at two MeV. As in all ββ-decay experiments, background suppression is of the utmost importance. There are, as in other neutrino experiments, cosmic rays to worry about, and especially radioactive materials, which can even be generated by cosmic rays

(Avignone 1988). The germanium experiment should be particularly well suited to search for 0ν ββ decay. An energy spike like that due to two electrons stands out better from the continuous background than a continuous spectrum.

Tellurium 128 has been found to have a half-life of (7.7 ± 0.4) x 10^{24} years (Bernatowicz 1992)! In spite of this successful invasion of the domain of ultra slow processes, neutrinoless double-beta decay has as yet not been observed. By 1999 the half-life-limit on Ge76 had been pushed to

$$T_{1/2} > 5.7\,x\,10^{25} y, \quad (90\%\, C.\,L.) \;.$$

An interpretation in terms of a transition amplitude proportional to the electron-neutrino mass as given by eq. (210b) results in (Baudis 1999)

$$m_{\nu} < 0.20\, eV \quad (90\%\, C.\,L.) \;.$$

At this point it may be of rather general relevance to quote from Furry's original article: "According to the older theory it seemed certain that double beta-disintegration could never be capable of observation because of its extremely minute probability, but the Majorana theory indicates that this is by no means necessarily the case" (Furry 1939).

(For access to the plethora of suggestions made in connection with 0ν ββ-decay see, e.g., Avignone 1994 and Klapdor-Kleingrothaus 1999).

7.4 RELATING TRANSFORMATIONS IN DOUBLE-BETA DECAY TO NEUTRINO OSCILLATIONS

It is of some interest to determine how the ad hoc transformations from current to mass eigenstate neutrinos, introduced to suggest the possibility of neutrino oscillations, are related to the more formal transformations used in the analysis of the intermediate states of the hypothetical 0ν ββ-decay.

With this purpose in mind we first rewrite the eqs. (208a) in terms of eigenfields:

$$\psi_{Lk} = U_{kl}\nu_{Ll} = U_{kl}(1-\gamma_5)/2\ \nu_l^+ \ , \ k = 1-3,\ l = 1-6$$
$$= U_{kl}(1-\gamma_5)/2\ \nu_l^- \ . \tag{215}$$

We now replace any eigenfield associated with a negative mass with the corresponding eigenfield of opposite superscript. (For ψ_{Rk}' this necessitates multiplication by -1).

Take the case of Majorana current-neutrinos. The submatrix M of the mass matrix **M** will now itself contain only two nonzero submatrices, namely M_L and $M_R^\dagger$. Since these occupy the diagonal slots, the eigenvector matrix of M, called U, will also decompose into two matrices,

$$U = \begin{pmatrix} U^1 & 0 \\ 0 & U^2 \end{pmatrix} \tag{216}$$

where, following eq. (204a),

$$U^{1T} M_L U^1 = D_L \tag{217}$$

and similarly with the superscript 2, subscript R, and a dagger on M.

Therefore the expression of the left-handed current-neutrino (field) in terms of the mass eigenfield will simply be

$$\psi_{Lk} = U_{kl}^1(1-\gamma_5)/2\ \nu_l^+ \tag{218a}$$

where

$$\nu^+ = \begin{pmatrix} (U^1)^{-1}\psi_L + (U^1)^T\psi_R^c \\ (U^2)^{-1}\psi_L^{c\prime} + (U^2)^{-1}\psi_R{}' \end{pmatrix} \tag{218b}$$

from eqs. (204) and (206). The two null submatrices of U prevent the mixing of primed and unprimed neutrinos - in the expression for ν^+ as well as in the expression of ψ_L in terms of ν^+. The index *l* goes now effectively only from one to three, quite as in eq. (184).

Dirac current-neutrinos also have a submatrix M with just two nonzero submatrices, one in the top right, the other in the bottom left slot; see eq. (203). If we write the diagonalizing matrix

$$U = 1/\sqrt{2} \begin{pmatrix} U^1 & U^2 \\ U^3 & U^4 \end{pmatrix} \tag{219}$$

we obtain

$$U^T M U = 1/2 \begin{pmatrix} U^{3T} m\, U^1 + U^{1T} m^T U_3 & U^{3T} m\, U^2 + U^{1T} m^T U^4 \\ U^{4T} m\, U^1 + U^{2T} m^T U^3 & U^{4T} m\, U^2 + U^{2T} m^T U^4 \end{pmatrix} . \tag{220a}$$

A convenient choice consists of unitary submatrices, and diagonal $U^{3T} m\, U^1$ and $U^{4T} m\, U^2$. $U^T M U$ is manifestly diagonal if

$$U^1 = \pm U^2 \ \ with \ \ U^3 = \mp U^4 \ . \tag{220b}$$

Thus is repeated on the next smaller scale what we encountered when diagonalizing the entire matrix $\mathbf{M}$: for every eigenmass there is an equal and opposite one. So, for three flavors this case results in three in general independent eigenmasses.

Summarizing, the original expansion of the left-handed current neutrinos in terms of a number of mass eigenstates equal to the number of flavors, eq. (184), is exact if used separately for Dirac and for Majorana current-neutrinos. What about the oscillation products? We discussed the "flavor oscillations", and did so in detail for the two component case, starting with eqs. (186). Flavor oscillations only violate flavor conservation; e.g., electron-neutrino tau-neutrino oscillations, as perhaps in the Sun (Valle 1999), or electron-antineutrino muon-antineutrino oscillations, as presumably upon fission.

With Dirac current-neutrinos the mass eigenstates are superpositions of primed and unprimed neutrinos, and thus oscillations of left-handed neutrinos will include the left-handed antiparticles of the right handed current neutrinos: eqs. (219) and (220) give us the eigenstate

$$\nu_k^+ = U_{kl}^{1\dagger}\,\psi_{Ll} + U_{kl}^{3\dagger}\,\psi_{Ll}^{c}{}' + U_{kl}^{1T}\,\psi_{Rl}^{c} + U_{kl}^{3T}\,\psi_{Rl}{}' \,,\; k,l = 1-3 \;. \tag{221a}$$

But there is also the eigenstate ν_{k+3}^- which has the same mass as ν_k^+ (ν_{k+3}^+ has the opposite mass of ν_k^+). In terms of the current neutrinos it is

$$\nu_{k+3}^- = U_{kl}^{1\dagger}\,\psi_{Ll} - U_{kl}^{3\dagger}\,\psi_{Ll}^{c}{}' - U_{kl}^{1T}\,\psi_{Rl}^{c} + U_{Rl}^{3T}\,\psi_{Rl}{}' \,,\; k,l = 1-3 \;. \tag{221b}$$

Thus an expansion as given by eq. (184) will not contain antineutrinos. For consider a left-handed current neutrino,

$$\begin{aligned} \psi_{Ll}(t) &= U_{lk}\,(1-\gamma_5)/2\;(\nu_k^+(t)\;\; or \;\; \nu_k^-(t)) \;,\; k = 1-6, l = 1-3 \\ &= (1-\gamma_5)/2\; U_{lk}^1(\nu_k^+(t) + \nu_{k+3}^-(t)) \;,\; k = 1-3 \\ & where \qquad \nu_k^+(t) = \nu_k^+\, exp(-iE_k^+ t) \;,\;\; no\, sum \end{aligned} \tag{222}$$

and similarly for $\nu_k^-(t)$. Substituting eqs. (221) in eq. (222), and remembering that $E_k^+ = E_{k+3}^-$, eliminates the current antineutrinos. It should: we are considering a pure Dirac mass matrix. - In eq. (221a) it is indicated to associate ν_k^+ with a positive mass.

Massive Majorana neutrinos should also be able to oscillate into antineutrinos and thus violate lepton-number conservation, although, as we can surmise from our introductory remarks, with an amplitude proportional to m/E (Bahcall 1978). We can already tell from the Lagrangian mass density, that Majorana neutrinos do not conserve lepton number; the mass density is not invariant under global gauge transformations.*

*In this connection a brief discussion of the number of CP-violating phases in mixing matrices for Dirac and Majorana neutrinos, and of their observability, should be mentioned. Bilenky 1980. Compare with the quark case mentioned at the end of Section 7.2.

7.5 SOME MASS GENERATING MECHANISMS FOR NEUTRINOS

Calling on the Higgs field, the formally simplest way to generate a neutrino mass is to proceed as in the standard model with the electron mass. One writes an amplitude

$$a = -ig'(\bar{l}_L \tau \Phi^* \psi_R) + h.c. \tag{223}$$

with τ the metric spinor in isospace, turning the Higgs doublet Φ^* into a properly transforming quantity, and ψ_R the wave function of a right-handed neutrino. As before, l_L is the left-handed lepton doublet. As to the bar, recall the definition $\bar{\psi} = (-i\beta\psi)^\dagger$. Equating the charged Higgs to zero and the neutral to the vacuum-expectation-value, v, results in the neutrino mass $m = g' v/\surd 2$.

As mentioned towards the end of chapter six, an isoscalar neutrino, if fitted into the standard model, would not be coupled to the neutral vector-boson, Z. It would only be coupled to the gravitational field. It is a so called sterile neutrino. Straightforward symmetry suggests three flavors for this neutrino. However, present day astrophysical arguments, more often than not based on the primordial deuterium to hydrogen ratio, favor a limit of less than five on the number of light neutrino kinds (Rev. of Particle Phys. 1998).

One mass generating mechanism for the left-handed two-component neutrinos, without invoking a sterile neutrino, involves a triplet Higgs field (Schechter 1980, Cheng 1980, Gelmini 1981. For other models see, e.g., Zee 1980 and Babu 1988). the amplitude is

$$a = ig''(\bar{l}_L^{\,c} \tau \boldsymbol{\tau} \cdot \boldsymbol{\varphi}\, l_L) + h.c.\,; \tag{224}$$

τ now multiplies a lepton doublet. In the form of the second-rank spinor of eq. (224) thc Higgs triplet is

$$\boldsymbol{\tau} \cdot \boldsymbol{\varphi} = \begin{pmatrix} H^+/\surd 2 & H^{++} \\ H^o & -H^o/\surd 2 \end{pmatrix} .$$

The triplet would also be coupled to the vector bosons, thus being partly responsible for their masses, and to the Higgs doublet (Schechter 1982. See also Grimus 1999).

In oscillation experiments the masses appear as differences of squared masses. Consequently, they will have to satisfy $\Delta m_{ij}^2 + \Delta m_{jk}^2 + \ldots + \Delta m_{li}^2 = 0$, where the number of terms is equal to or less than the number of flavors. To judge by the two favored Δm^2 values following from the solar data, the favored Δm^2 from the atmospheric data and the minimum Δm^2 from the Los Alamos data, three flavors will not do. The solar values are much smaller than the other two, which in turn are not close to one another. An atmospheric Δm^2 one order of magnitude bigger than the favored value approaches the minimum Los Alamos value (see p. 121). But the latter is hardly realistic. Invoking a fourth flavor goes against results from Z decays. Subtracting the sum of the measured partial widths of the Z decay into charged leptons and quarks from the total width, and using the standard-model value of very close to two for the ratio of the partial widths of the decay into a neutrino and the decay into its charged lepton partner, leaves a width corresponding to just about exactly three kinds of neutrinos (Mnich 1999). This result comes from four experiments at the Large-Electron-Positron collider at CERN, "the LEP experiments". The sterile neutrino is being called to the rescue.

APPENDIX

1. A "group" is a totality of objects called "elements", between which a law of composition is defined, called "multiplication". Thus the "product" of two elements is also an element.

The elements and their multiplication satisfy three conditions:

1) (AB)C = A(BC)

i.e., the associative law holds.

2) There is one and only one "unit", E, such that for any element A

AE = EA = A

3) Every element has a unique "reciprocal", i.e.,

BA = E

Then AB = E follows.

2. A matrix representation of a group is "irreducible" if at least one matrix has only one block along the left top to bottom right diagonal, which block cannot be reduced (by a similarity transformation) into smaller blocks. Every unitary representation can be fully reduced. Only a constant multiple of the unit matrix commutes with all elements of an irreducible representation.

3. For a discussion of the inhomogeneous Lorentz group, L_i, it is useful to introduce the operator (matrix)

$$\Gamma_\mu = (1/2)\,\varepsilon_{\mu\nu\omega\chi} J^{\nu\omega} p^\chi \tag{A1}$$

where ε is the totally antisymmetric "Levi-Civita tensor", which is here standardized by setting

$$\varepsilon_{1230} = +1 \ . \tag{A2}$$

Apparently

$$\Gamma^\mu p_\mu = 0 \ .$$

The scalar product $\Gamma^\mu\Gamma_\mu$ commutes with all the generators of L_i. So does of course $p^\mu p_\mu$. Summing over ν, ω and χ and expressing the angular-momentum-tensor in terms of **J** and **K** (see section 2), results in

$$\boldsymbol{\Gamma} = \mathbf{J}p^0 + \mathbf{K}x\mathbf{p}\,,\;\; \Gamma^0 = \mathbf{J}\cdot\mathbf{p}\;. \tag{A3}$$

With existing rest frame we see that

$$\Gamma^\mu \Gamma_\mu = m^2 J(J+1)\;. \tag{A4}$$

Consider subspaces with

$$p_1 = p_2 = 0\;.$$

Then $\Gamma^\mu \Gamma_\mu$

$$= (J_1 p^0 + K_2 p^3)^2 + (J_2 p^0 - K_1 p_3)^2 + (J_3 p^0)^2 - (J_3 p_3)^2 \tag{A5}$$

is the effective invariant. For $p_3 = p^0$

$$\Gamma^\mu \Gamma_\mu = ((E_1)^2 + (E_2)^2)(p^0)^2 \tag{A6a}$$

$$E_1 = K_1 - J_2 \;\; and \;\; E_2 = K_2 + J_1\;. \tag{6b}$$

$E_1, E_2\, and\, J_3$ obey the commutation relations of the generators of the two dimensional Euclidian group, E_2 (see sections 3.2.1 and 4.1). Restriction to subspaces effecting $E_1 = E_2 = 0$ leads to

$$\Gamma^\mu = \Lambda p^\mu\;. \tag{A7}$$

This is so because we can now write

$$\Gamma^{\mu}\Gamma_{\mu} = 0 \tag{A8}$$

generalizing to an arbitrary direction from that of the third axis. And since we already have $\Gamma^{\mu}p_{\mu}$ and $p^{\mu}p_{\mu}$ equal to zero, the proportionality as given by eq. (A7) follows. Λ is the diagonal matrix made up of eigenvalues λ of the rotation generator $\mathbf{J}\cdot\mathbf{p}/|\mathbf{p}|$. This generator is obviously one of the three generators of the two-dimensional Euclidian group, see above. With the other two represented trivially, let us find the λ values as they relate to A and B of a representation (A,B) of the homogeneous Lorentz group. The subspaces $\chi(\lambda,\theta=0,\varphi)$ and $w(\lambda,\theta=0,\varphi)$, see eqs. (50) and (75), respectively, behave according to

$$(\varepsilon_1 E_1 + \varepsilon_2 E_2 + \varepsilon_3 J_3)\chi = \varepsilon_3 \lambda X \tag{A9}$$

$$\begin{aligned} \textit{where} \quad E_1 &= K_1 - J_2 = iB_1 - B_2 - iA_1 - A_2 \\ E_2 &= K_2 + J_1 = B_1 + iB_2 + A_1 - iA_2 \, . \end{aligned} \tag{A10}$$

$$\textit{and} \quad X = \chi \textit{ and } w \textit{ , respectively} .$$

With the "magnetic quantum numbers" m and n, where $A = m_{\max}$ *and* $B = n_{\max}$, eqs. (A9) and (A10) give

$$(A_3 + B_3)_{mn,m'n'} X_{m'n'} = \lambda X_{mn} \tag{A11a}$$

$$(A_1 - iA_2)_{mn,m'n'} X_{m'n'} = 0, \; (B_1 + iB_2)_{mn,m'n'} X_{m'n'} = 0 \, . \tag{A11b}$$

We are dealing here with the standard raising and lowering operators of nonrelativistic quantum mechanics:

$$(J_1 + iJ_2)_{m+1,m} = (J_1 - iJ_2)_{m,m+1} = \sqrt{((j-m)(j+m+1))} \tag{A12}$$

all the other matrix elements being zero. Thus

$$X_{m,n} \neq 0 \ \ only\ for\ m = -A, n = B \ . \tag{A13a}$$

The third equation then yields

$$\lambda = B - A \ . \tag{A13b}$$

Thus for a massless particle species of helicity λ, representations of L_h are restricted to those satisfying the last equality (Weinberg 1965).

4. A matrix with elements a_{kl} shall operate on a set of vectors x_m with components x_{lm}. The vectors are eigenvectors with eigenvalues λ_m if

$$a_{kl}x_{lm} = \lambda_m x_{km} \qquad no\ sum\ over\ m\ .$$

The indices go from l to n. We write

$$\begin{aligned} a_{kl}x_{lm} &= x_{kl}\delta_{lm}\lambda_m \\ x^{-1}_{jk}a_{kl}x_{lm} &= \delta_{jm}\lambda_m \ . \end{aligned} \tag{A14}$$

5. It was suggested in section 3.1.4 to use the 2-spinors ξ_{As} and ξ_{Bs} to calculate the longitudinal polarization of β particles. Here we shall use the 2-spinors χ_{As} and χ_{Bs} given by eqs. (56), i.e., the helicity representation.

$$\begin{aligned}<\boldsymbol{\sigma}\cdot\mathbf{p}>_{As} &= \chi_{As}^{*}\,\sigma\cdot p\,\chi_{As} = -E\chi_{As}^{*}\chi_{As} + m\chi_{As}^{*}\chi_{Bs}\\ &= (-E\exp(-\alpha) + m)/N^2 \ \ for\ s = 1/2\\ &= (-E\exp(\alpha) + m)/N^2 \ \ for\ s = -1/2\\ N^2 &= \exp(\alpha) + \exp(-\alpha)\\ &= 2E/m\ .\end{aligned}$$

Consequently,

$$\begin{aligned}<\boldsymbol{\sigma}\cdot p>_{A1/2} &= m^2p/2E(E+p)\\ <\boldsymbol{\sigma}\cdot p>_{A-1/2} &= -p(E+p)/2E\\ <\boldsymbol{\sigma}\cdot p>/p &= -p/E\ .\end{aligned} \tag{A15}$$

Note the mass dependence of the first expectation value. Here we used E for the energy.

6. To obtain the Dirac equation with the electromagnetic interaction from a Lagrangian one must add the term $e\,\bar{\psi}\gamma^{\mu}\psi A_{\mu}$ to the Lagrangian density as given by eq. (174). Since a complete solution of the Euler-Lagrange-equations will also involve the differentiation with respect to A^{μ}, the addition to the Lagrangian density of $-F^{\mu\nu}F_{\mu\nu}/4$ is indicated. Thus, the formalism will also result in the inhomogeneous Maxwell equations. The conserved-current-density of eqs. (114) is forced upon us.

7. The development of section 5.2 can be continued: in the integrand of the first term on the right of eq. (132) the product $exp(\pm i\varphi(l,p^{\kappa}))\,a_{\pm 1}(p^{\nu})$ shall be called $\surd(p'^{0}/p^{0})\,a_{\pm 1}'(p'^{\nu})$, where $p'^{\nu} = l^{\nu}{}_{\mu}p^{\mu}$.

With no invariant sum over s involved, a^{s} may be replaced by a_{s} . Consequently said first term will be

$$\begin{aligned}(2\pi)^{-3/2}\int(d^3p/2p^0)\surd(2p'^0)\,\wp_{\pm 1}(p'^{\nu})(a_{\pm 1}'(p'^{\nu})\exp(ip^{\mu}x_{\mu})\\ +\,a_{+1}^{*}{}'(p'^{\nu})\exp(-ip^{\mu}x_{\mu})\,.\end{aligned} \tag{A18a}$$

The differential d^3p/p^0 (not $\surd p^0$!) is a Lorentz invariant:

$$dp/p^0 = (dp' + v\,dp'^0)/(p'^0 + vp')$$

$$and\ p^0 dp^0 = p\,dp\ .$$

So is $p^\mu x_\mu$. Therefore expression (A18a) becomes

$$(2\pi)^{-3/2} \int (d^3p/\sqrt{2p^0})\, \wp_{\pm 1}(p^\nu)(a_{\pm 1}'(p^\nu) \exp(ip^\mu x_\mu') + \ldots)\ . \qquad \text{(A18b)}$$

We have come to a jumping off point for quantum field theory. The Fourier coefficients become operators and the primed are related to the unprimed ones of the same argument by

$$a_{\pm 1}'(p^\nu) = U^{-1}(l)\, a_{\pm 1}(p^\nu)\, U(l)\ . \qquad \text{(A19)}$$

So eq. (132) becomes (Weinberg 1965)

$$l^\mu_\nu A^\nu_{\pm 1}(x^\kappa) = U^{-1}(l)\, A^\mu_{\pm 1}(x'^\kappa)\, U(l) + gradient\ terms\ . \qquad \text{(A20)}$$

For a massive potential simply substitute Φ for A in the last equation and drop the gradient terms. The demonstration is left to the readers.

8. At the end of our brief discussion of the electroweak theory we encountered the Lagrangian density from which the Dirac equation for a free particle - or a free field, in a classical sense - can be derived. Now to the Lagrangian density of a 4-vector field. For a massive field it is

$$\mathcal{L} = -1/4\, F^{\mu\nu}(x^\kappa) F_{\mu\nu}(x^\kappa) - m^2/2\, \Phi^\mu(x^\kappa)\, \Phi_\mu(x^\kappa)$$
$$F^{\mu\nu} \equiv \partial\Phi^\nu/\partial x_\mu - \partial\Phi^\mu/\partial x_\nu\ . \qquad \text{(A21)}$$

The Euler-Lagrange equations, see eqs. (173), for Φ^μ are then

$$\nabla \times \mathbf{B} - \dot{\mathbf{E}} = -m^2 \mathbf{\Phi} \; and \; \nabla \cdot \mathbf{E} = -m^2 \Phi^0 \; . \tag{A22}$$

where we borrowed the designations B and E from electromagnetism. As before,

$$E^k = F^{0k} \; and \; B^k = F^{lm} \; .$$

With the canoncially conjugate (so called) momentum

$$\pi^k(x^\kappa) = \partial \mathcal{L} / \partial \dot{\mathbf{\Phi}}_k(x^\kappa) = -E^k(x^\kappa) \tag{A23}$$

and $\pi^\aleph(x^\kappa)$ equal to zero because $\mathcal{L}$ does not depend on Φ_0, compare with section 5.1., the Hamiltonian density

$$\mathcal{H} = \mathbf{\Phi}_\mu \, \partial \mathcal{L} / \partial \dot{\mathbf{\Phi}}_\mu - \mathcal{L} \tag{A24a}$$

is evaluated as

$$\mathcal{H} = -\mathbf{E} \cdot \mathbf{\Phi} + (B^2 - E^2)/2 + m^2(\Phi^2 - (\Phi^0)^2) \; . \tag{A24b}$$

The first two terms equal

$$(B^2 + E^2)/2 + \mathbf{E} \cdot \nabla \Phi^0 \; .$$

Let us calculate the Hamiltonian. We already know the expansion of $\Phi^\mu(x^\lambda)$, see eqs. (101) to (104), in terms of standard vectors, Fourier coefficients and plane waves. (For a different view of the expansion see Ahrens 1984). The calculation is simplified, if p_1 and p_2 are at first set equal to zero:

$$\Phi^\mu(x^\lambda) = (2\pi)^{-3/2} \int (dp/\surd(2p^0))\, B^\mu_\nu \rho^\nu_s (a^s \exp(ip^\kappa x_\kappa) + a^{-s*} \exp(-ip^\kappa x_\kappa)) \ . \tag{A25}$$

For a particular p and p^0 the transverse B and E components are proportional to p, and p^0, respectively. The longitudinal B is zero, while the longitudinal E is proportional to $p^0 \cosh\alpha - p \sinh\alpha = m$. Furthermore,

$$\mathbf{E}\cdot\nabla\Phi^0 = \nabla\cdot\Phi^0\mathbf{E} - \Phi^0\nabla\cdot\mathbf{E} \ .$$

It is left to the reader to show, that the only noncancelling contributions to

$$H = \int \mathcal{H}\, d^3x \tag{A26a}$$

come from the terms containing the products

$$a_k(p)\, a^*_k(p')\exp(i(p-p')x), \textit{ no sum over } k \ .$$

One obtains

$$\begin{aligned} &= \int (dp/2p^0)[(p^2+(p^0)^2)(a_1a_1^* + a_2a_2^*) + (p^0\cosh\alpha - p\sinh\alpha)^2 a_3a_3^* \\ &+ (2m^2\sinh^2\alpha)a_3a_3^* + m^2(a_1a_1^* + a_2a_2^* + (\cosh^2\alpha - \sinh^2\alpha)a_3a_3^*)] \\ &= \int p^0 a_k a_k^*\, dp \ . \end{aligned} \tag{A26b}$$

In the above $a_k(p)$ is abbreviated to a_k. Dropping the condition $p_1 = p_2 = 0$ results in the last integral to be over d^3p .

9. Classical field theory, like classical mechanics, may be deduced from an action principle,

$$\delta \int \mathcal{L}\, dx = 0 \; . \tag{A27a}$$

In the field theory, $\mathcal{L}$ is the "Lagrangian density",

$$\mathcal{L} = \mathcal{L}(\psi_a(x)\,,\; \frac{\partial \psi_a}{\partial x}) \; . \tag{A27b}$$

The field functions $\psi_a(x)$ are the dynamical variables of the theory, corresponding to the $q_k(t)$ of particle mechanics. Whereas the position variables of mechanics have an index, here *k*, which assumes the integer values labeling the degrees of freedom of a particle system, the field variables have the "continuous index" **r**, the position, at which ψ_a is evaluated at some time. Thus the sum over *k* in mechanics corresponds to the 3-space integration in field theory; x is the collective of all four x^{μ}, $\mu = 0$ to 3, and dx denotes the volume element of 4-space.

The equations of a field theory usually depend on space derivatives in addition to the time derivative. So, proceeding analogously to the well known derivation in classical mechanics, the differential equations obtained from eq. (A27a) are

$$\frac{\partial \mathcal{L}}{\partial \psi_a} - \frac{\partial}{\partial x^{\nu}} \frac{\partial \mathcal{L}}{\partial \dfrac{\partial \psi_a}{\partial x^{\nu}}} = 0 \; . \tag{A28}$$

They are also called the Euler-Lagrange equations, since they clearly constitute the straightforward generalization from the one parameter case of mechanics to the four parameter case of field theory. To arrive at these equations, it is necessary to let the variations $\delta\psi_a(x)$ vanish not only at the two endpoints of the time but also at the boundary of the 3-space integration.

In describing a given physical system, two observers, whose coordinates are connected by a symmetry transformation, must be able to use the same form for the action:

$$\Delta A = \int \mathcal{L}(\psi'_a(x'), \frac{\partial \psi_a'}{\partial x'^\nu}) dx' - \int \mathcal{L}(\psi_a(x), \frac{\partial \psi_a}{\partial x^\nu}) dx = 0 . \tag{A29}$$

We restrict ourselves to continuous transformations, and consider the infinitesimal case. To first order one expands the field functions as follows:

$$\psi_a'(x') = \psi_a(x) + (\psi_a'(x) - \psi_a(x)) + \frac{\partial \psi_a}{\partial x^\nu}(x'^\nu - x^\nu) .$$

having in the last term replaced ψ_a' by ψ_a. Furthermore, the difference of the volume elements in 4-space is rather obviously

$$\delta(dx) = \frac{\partial \delta x^\nu}{\partial x^\nu} dx$$

where the coordinate transformation has been expressed as

$$x'^\nu = x^\nu + \delta x^\nu . \tag{A30a}$$

Eq. (A29) now becomes

$$\int \frac{\partial}{\partial x^\nu} \left(\frac{\partial \mathcal{L}}{\partial \frac{\partial \psi_a}{\partial x^\nu}} \delta \psi_a + \mathcal{L} \delta x^\nu \right) dx = 0 \tag{A31}$$

where $\delta\psi_a$ is defined as the "substantial variation"

$$\delta \psi_a \equiv \psi_a'(x) - \psi_a(x)$$

also referred to as the "change in form".

With both, the coordinate change $\delta x^{\nu}$and the "local variation" $\psi_a'(x') - \psi_a(x)$ given in terms of the infinitesimal parameters of the transformation, one is in possession of a conservation law. This is the meaning of the Noether Theorem.

The derivation of eq. (A31) is facilitated, if one draws on the following relation between local and substantial variations,

$$F(f'(x')) - F(f(x)) = F(f'(x)) - F(f(x)) + \frac{dF}{dx}\,\delta x\ .$$

The preceding may become more intuitive as one recalls the analogous situation in classical mechanics:

$$\int L(q_k'(t'), \dot{q}_k'(t'))dt' = \int [L(q_k(t), \dot{q}_k(t)) + \frac{d(\delta F)}{dt}]\,dt \qquad \text{(A32a)}$$

$$with \quad \delta F = p_k\,\delta q_k + L\,\delta t$$

where p_k is the canonically conjugate momentum of q_k . In the case of a symmetry transformation one has for the infinitesimal generating function, δF,

$$\frac{d(\delta F)}{dt} = 0\ . \qquad \text{(A32b)}$$

An important, perhaps obvious difference between coordinate transformations in these two areas of physics is to be noted, however: in mechanics the infinitesimal translation of a Cartesian coordinate,

$$x_k' = x_k + \varepsilon \qquad \text{(A33)}$$

is generated by $p_k \equiv \partial L/\partial \dot{x}_k$, a fact frequently expressed in terms of the Poisson bracket of p_k with an arbitrary function of the canonical variables,

$$\{p_k, f(x_l, p_m)\} = -\frac{\partial f}{\partial x_k} . \tag{A34}$$

In field theory the translation of the field point variable x_k leads to the change in form of the field function

$$\begin{aligned} \delta\psi_a &\equiv \psi_a'(x) - \psi_a(x) \\ &= \psi_a'(x') - \psi_a(x) - \frac{\partial\psi_a}{\partial x^k}\varepsilon \\ &= -\frac{\partial\psi_a}{\partial x^k}\varepsilon . \end{aligned} \tag{A30b}$$

Now consider eq. (A31). Let the field functions behave at spatial infinity in such a manner as to guarantee the vanishing of the surface integral. The remainder may be written

$$\int (\pi^a\,\delta\psi_a + \mathcal{L}\,\delta x^0)\,dv\Big|_{t_1}^{t_2} = 0 \tag{A35}$$

with π^a, the "canonically conjugate momentum" of ψ_a, defined by

$$\pi^a \equiv \frac{\partial\mathcal{L}}{\partial\dfrac{\partial\psi_a}{\partial t}} . \tag{A36}$$

For the case of the translation the integrand becomes

$$\delta I = -\pi^a\frac{\partial\psi_a}{\partial x_k}\varepsilon$$

and the field theoretic generating function

$$\delta F = \int (\delta I) dv \equiv p_k \varepsilon \ . \tag{A37a}$$

The "physical analogue" of eq. (A34) is then

$$\{p_k, \psi_a(x)\} = \frac{\partial \psi_a}{\partial x^k} \ . \tag{A38}$$

There is no minus on the right of this equation; it is not a "mathematical analogue" of eq. (A34). The mathematical analogue of eq. (A34) reads

$$\{\pi^a(\mathbf{r},t), f[\psi_b(\mathbf{r}',t), \ldots]\} = -\frac{\delta f}{\delta \psi_a(\mathbf{r},t)} \ . \tag{A39}$$

The square brackets mean "functional of". Remember, in field theory the positions are continuous indices. That explains the functional derivative on the right side of eq. (A39).

To complete the set of 4 dimensional translation generators, i.e., the generators in the sense of eq. (A38), one adds to the linear momenta, p_k, the time translation generator,

$$p_0 = -H \equiv \int (\mathcal{L} - \pi^a \dot{\psi}_a) dv \tag{A37b}$$

Remember,

$$\psi_a'(x^0) = \psi_a(\tilde{x}^0)$$

and thus

$$\delta \psi_a = -\varepsilon \partial \psi_a / \partial x^0 \ .$$

Furthermore, $\delta x^0 = \varepsilon$. With a Poisson bracket we have

$$\left\{H, \psi_a(x)\right\} = -\frac{\partial\psi}{\partial x^0} \ . \tag{A40}$$

We shall conclude this brief discussion of Noether's theorem with a look at an "internal symmetry transformation," the local gauge transformation of electrodynamics. Since no coordinate change occurs, we immediately move on to the transformation of the integrand of the action, the Lagrangian density. Infinitesimally we then have for the "matter field" ψ (as distinguished from the electromagnetic field)

$$\psi(x) \to (1 + ie\,\varphi(x))\,\psi(x) \tag{A41}$$

and so for $\mathcal{L} = \mathcal{L}_m(\psi, \partial\psi/\partial x^\mu)$

$$\begin{aligned}\delta\mathcal{L} = ie\varphi(x)&\left(\frac{\partial\mathcal{L}}{\partial\psi}\psi + \frac{\partial\mathcal{L}}{\partial(\partial\psi/\partial x^\mu)}\frac{\partial\psi}{\partial x^\mu} - \frac{\partial\mathcal{L}}{\partial\psi^*}\psi^* - \frac{\partial\mathcal{L}}{\partial(\partial\psi^*/\partial x^\mu)}\frac{\partial\psi^*}{\partial x^\mu}\right)\\ &+ ie\left(\frac{\partial\mathcal{L}}{\partial(\partial\psi/\partial x^\mu)}\psi - \frac{\partial\mathcal{L}}{\partial(\partial\psi^*/\partial x^\mu)}\psi^*\right)\partial\varphi/\partial x^\mu \ .\end{aligned} \tag{A42}$$

For constant φ, certainly a permissable choice, the demand that $\mathcal{L}$ be symmetric under the transformation (A41) yields a conserved current, j^μ. This is of course provided, $\mathcal{L}_m$ satisfies the Euler-Lagrange equation, as the reader will notice when looking at the expression in the first parenthesis of eq. (A42). The case of a coordinate dependent gauge function then leads to a change in $\mathcal{L}$.

This change can be nullified by the presence of an interaction density $\mathcal{L}_i$, such that

$$\frac{\partial\mathcal{L}_i}{\partial A_\mu}\frac{\partial\psi}{\partial x^\mu} = -\,ij^\mu\frac{\partial\psi}{\partial x^\mu}$$

or

$$\frac{\partial\mathcal{L}_i}{\partial A_\mu} = -\,ij^\mu \ . \tag{A43}$$

We see that the change in the potential functions A^μ accompanying the change in ψ is

$$A^{\mu} \rightarrow A^{\mu} + \partial\varphi/\partial x_{\mu} \qquad \text{(A44)}$$

the original gauge transformation.

With vanishing current density at infinity, conservation of charge follows:

$$e \int j^0 d^3x = constant \ . \qquad \text{(A45)}$$

In section 5.2 we saw that a Lorentz transformation of the radiation potential results in a gauge transformation. The mathematical reason for this is the choice of helical polarization vectors: once expressed as functions of a 4-vector of norm zero, the 4-momentum of mass zero, the polarization vectors transform under E(2) in such a way that nonzero momentum component and energy stay the same. But the polarization states of the radiation potential are transverse or left and right circularly polarized, as made apparent by the nonzero Fourier coefficients. Thus the gradient terms, as shown in section 5.2.

Clearly, all this involves plane waves, i.e., free fields or free particles. Nevertheless, for this special case it can then be said that charge conservation follows from Lorentz invariance. This is of particular importance in the quantum field theory, where the usual perturbative approach involves the "interaction representation", which is built on free fields (Weinberg 1996).

10. A vector is said to be contravariant, if its components transform like coordinate differentials,

$$d\xi'^k = \frac{\partial \xi'^k}{\partial \xi^\ell} d\xi^\ell \, . \qquad \text{(A46a)}$$

Clearly, this definition is quite general. The transformation coefficient

$$t^k_{\ \ell} = \partial\xi'^k / \partial\xi^\ell \qquad \text{(A46b)}$$

may depend on the coordinates. A vector is said to be covariant, if its components transform like gradient components

$$\frac{\partial}{\partial \xi'^k} = \frac{\partial \xi^\ell}{\partial \xi'^k} \frac{\partial}{\partial \xi^\ell} \ . \tag{A47a}$$

The transformation coefficients are now

$$t^{-1\ell}{}_k = \partial \xi^\ell / \partial \xi'^k \ . \tag{A47b}$$

Local gauge transformations are coordinate dependent. But in this text we are only interested in their application to internal spaces, such as the weak isospace. Lorentz and particle state transformations are coordinate independent.

The transformation laws $\varphi'^k = t^k{}_\ell \varphi^\ell$ *and* $\varphi_k' = \varphi_\ell t^{-1\ell}{}_k$ are extended to spinors as well as to multi component objects of higher rank.

In the beginning of section 4.2 we juxtaposed the transformation properties of components and basis vectors. The covariant basis vectors transformed according to

$$V'_k = V_\ell t^\ell{}_k \ . \tag{A48}$$

We now look at a coordinate dependent function, $\psi(x^\mu)$, as a member of a set of contravariant basis vectors in Hilbert space, $\psi^k(x^\mu)$, and transform

$$\psi^k(x^\mu) \to \psi^k(\tilde{x}^\mu) = t^{-1k}{}_\ell \psi^\ell(x^\mu) \ . \tag{A49a}$$

The covariant basis vectors will consequently transform like

$$\psi_k(x^\mu) \to \psi_k(\tilde{x}^\mu) = \psi_\ell(x^\mu) t^\ell{}_k \tag{A49b}$$

We then have for the components in an expansion $\varphi^k \psi_k(x^\mu)$

$$\varphi^k \rightarrow \varphi'^k = t^k{}_\ell \varphi^\ell \ . \tag{A49c}$$

Here we are not talking about multi component functions like the 4-potential or the 4-spinor, but rather a set of coordinate dependencies with only the coordinates transforming. E.g., consider the set of five second order spherical harmonics and transform $x^k \rightarrow \tilde{x}^k$. $\psi^k(x^\mu) \rightarrow \psi'^k(x^\mu) = \psi^k(\tilde{x}^\mu)$, $k = 1$ *to* 5 follows (see eq. (1b).

11. An n-by-n matrix has n^2 elements, and thus before any conditions $2n^2$ parameters, since the elements are in general complex. 1), unitarity imposes n conditions of type $\sum_{k=1}^{n} u_{kl} u_{kl}^* = 1$, *no sum over* l; 2), it imposes n(n-1) conditions of type $\sum_{k=1}^{n} u_{kl} u_{km}^* = 0, l \neq m$. There are n(n-1)/2 such independent sums, but, because the products are in general complex, an additional factor of two appears. A special unitary matrix has thus n^2 - 1 independent parameters. (Its determinant is plus one).

From nonrelativistic quantum mechanics we are familiar with SU(2), the group representable by two-by-two special unitary matrices. These matrices may be generated by the Pauli sigmas divided by two.

Our approach to the group SU(3) shall be brief, and therefore physical. Any satisfactory purely mathematical approach is necessarily somewhat lengthy (Behrends 1962, Fronsdal 1963).

Consider three particles, e.g., proton, neutron and lambda - reminiscent of the erstwhile Sakata model (Sakata 1956) - or up-quark, down-quark and strange-quark. We shall use the latter example and thus have three states with the I, I_3 and Y assignments

$$\begin{aligned} I = 1/2, I_3 = \pm 1/2, Y = 1/3 \ \ and \\ I = 0, I_3 = 0, Y = -2/3 \ . \end{aligned} \tag{A50}$$

Transitions between all states shall be possible, giving rise to six ladder operators, in generalization of the two for I-spin by itself. Diagonal operators are at first three in number, but elimination of the unit reduces them to two. We see at this point that identification of one of the two diagonal operators with I_3 , which together with the ladder operators $I_1 \pm iI_2 = I^\pm$ forms the generators of the subgroup SU(2), is mathematically quite arbitrary (see section 5.6).

The I-spin ladder operators connect the I = ½ states. Adhering to a rather common convention, the combination $I_4 \pm iI_5 = V^\pm$ shall connect the I = 0 with the $I_3 = ½$ state, and $I_6 \pm iI_7 = U^\pm$ the I = 0 with the $I_3 = - ½$ state. If we arrange the three states in the column

$$q = \begin{pmatrix} u \\ d \\ s \end{pmatrix} \tag{A51}$$

with obvious meaning, then the matrices I_k, k = 1, 2 and 4 to 7, multiplied by two, are in that order

$$\begin{pmatrix} 0 & 1 & 0 \\ 1 & 0 & 0 \\ 0 & 0 & 0 \end{pmatrix}, \begin{pmatrix} 0 & -i & 0 \\ i & 0 & 0 \\ 0 & 0 & 0 \end{pmatrix}, \begin{pmatrix} 0 & 0 & 1 \\ 0 & 0 & 0 \\ 1 & 0 & 0 \end{pmatrix}, \begin{pmatrix} 0 & 0 & -i \\ 0 & 0 & 0 \\ i & 0 & 0 \end{pmatrix}, \begin{pmatrix} 0 & 0 & 0 \\ 0 & 0 & 1 \\ 0 & 1 & 0 \end{pmatrix}, \begin{pmatrix} 0 & 0 & 0 \\ 0 & 0 & -i \\ 0 & i & 0 \end{pmatrix}.$$

The diagonal generators are

$$I_3 = \frac{1}{2}\begin{pmatrix} 1 & 0 & 0 \\ 0 & -1 & 0 \\ 0 & 0 & 0 \end{pmatrix} \; and \; Y = \frac{1}{3}\begin{pmatrix} 1 & 0 & 0 \\ 0 & 1 & 0 \\ 0 & 0 & -2 \end{pmatrix}.$$

From the theory of angular momentum we are accustomed to the commutation relation

$$[I^+, I^-] = 2I_3 \; . \tag{A52a}$$

To it we can now add the relations

$$\begin{aligned} [U^+, U^-] &= 2U_3 \equiv 3Y/2 - I_3 \; and \\ [V^+, V^-] &= 2V_3 \equiv 3Y/2 + I_3 \end{aligned} \tag{A52b}$$

Instead of the I_k, k = 1 to 3, we could use I_4, I_5 and V_3 or I_6, I_7 and U_3 as generators of the subgroup SU(2). The corresponding "hypercharge" would be

$$Y_U = -I_3 - Y/2 \text{ and } Y_V = I_3 - Y/2$$

respectively. But for hadronic quarks, the I_k, k = 1 to 3, are clearly the generators of choice.

The commutation relations of all eight generators may be written

$$[I_k, I_l] = i f_{klm} I_m \qquad \text{(A53)}$$

The "structure constants" f_{klm} may be obtained with the help of the three-by-three matrices representing the eight I_k. They are antisymmetric in all three indices, except when one of these has the value eight. To eliminate this marr, one multiplies the hypercharge matrix by √3/2. Mathematical derivations usually end up with such an I_8 right away.

Just as the completely antisymmetric structure constants δ_{klm} given by

$$[I_k, I_l] = i\,\delta_{klm} I_m$$

form a matrix representation of SU(2), so do the completely antisymmetric f_{klm} form a representation of SU(3). These are the "regular representations", three dimensional for SU(2) and eight dimensional for SU(3). They are also subspaces of their respective groups. The importance of these representations in particle physics was mentioned in the discussion of "isospin et cetera".

12. Exp(sL) is the "exponential map" of a continuous one-parameter group, with s the parameter and L the generator. Clearly, the group is Abelian:

$$\exp(sL)\exp(s'L) = \exp(s+s')L \ . \qquad \text{(A55)}$$

Let us generalize to a connected n-parameter group with generators L, L′, etc., and parameters s, s′, etc. In a "connected group" all elements are continuously connected to the unit element or "identity". Restricting ourselves to infinitesimal

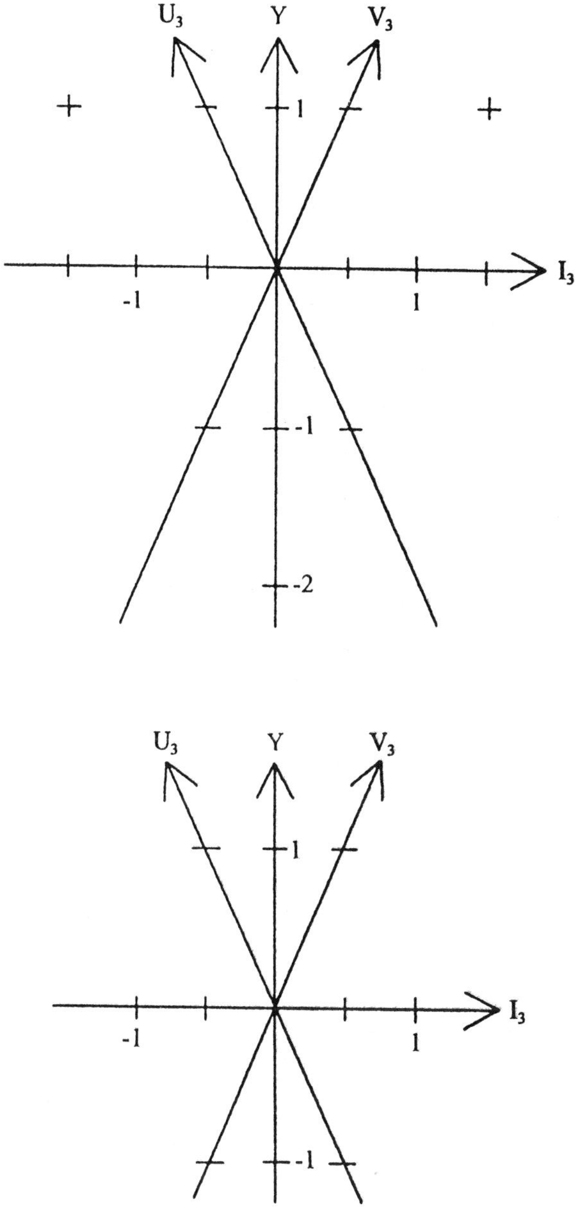

Figure 4: SU(3) decuplet and octet.

See pages 88 and 89.

parameters, the multiplication law to third order is

$$(1+X+X^2/2+X^3/6)\,(1+Y+Y^2/2+Y^3/6) = \\ (1+X+Y+(X+Y)^2/2+[X,Y]/2 \\ +(X+Y)^3/6+(X+Y)\,[X,Y]/4+[X,Y]\,(X+Y)/4+[[X,Y],Y-X]/12 \tag{A56}$$

where X = εL and Y = ε′L′. The right side of this equation is made up of the zeroth to third order terms in X and Y of the sum

$$1+A+A^2/2+A^3/6$$

where A = X + Y + [X,Y]/2 + [[X,Y], X-Y]/12. Including all orders, the equation is the "Baker-Campbell-Hausdorff formula"

$$\exp X \exp Y = \exp \sum_n c_n(X,Y) \tag{A57}$$

where the first few c_n are

$$c_0 = 0,\ c_1 = X+Y,\ c_2 = [X,Y]/2 \\ c_3 = [[X,Y],\ Y-X]/12 \\ c_4 = -([Y,[X,[X,Y]]] + [X,[Y,[X,Y]]])/48\ .$$

The X and Y need of course no longer be infinitesimal. (For the recursion relation for c_{n+1} and a discussion of convergence, see Varadarajan 1974).

The formula is quite general, but here we are only interested in its particular use as a group multiplication. Of prime importance is the appearance of commutators but beyond that of no novel features.

To understand the meaning of the commutator in the context of connected groups, it is useful to go back to the transformation law of N coordinates x_k involving n parameters s_r

$$x'_k = f_k(x_l; s_r) \tag{A58}$$

In the neighborhood of the identity we have, with $dx_k = x'_k - x_k$,

$$dx_k = (\partial f_k/\partial s_r)_o \, ds_r \tag{A59}$$

where the subscript zero denotes an evaluation of the derivative at all $s_r = 0$. There $f_k = x_k$. (Remember, unless noted otherwise, an index occurring twice in a product implies a sum.)

For linear transformations the partial derivative $(\partial f_k/\partial s_r)_o$ is a simple sum,

$$(\partial f_k/\partial s_r)_o = L^{\,k}_{r\,l} x_l \tag{A60}$$

(The vertical position of an index shall be of no significance). We have arrived at the generators L_r. On account of eq. (A59) the infinitesimal group multiplication becomes an addition. In the linear case this means

$$(1 + L_1 ds_1)(1 + L_2 ds_2) = 1 + L_1 ds_1 + L_2 ds_2 \ .$$

To get to the commutator of different generators, we consider the "group commutator"

$$g_1 g_2 g_1^{-1} g_2^{-1} = \Gamma$$

or the equation

$$g_1 g_2 = \Gamma g_2 g_1 \tag{A61}$$

With $g_1 \equiv g(s_1)$ and $g_2 \equiv g(s_2)$ we immediately see that $\Gamma(0,s_2) = \Gamma(s_1,0) = 1$, and thus in particular $\Gamma(0,ds_2) = \Gamma(ds_1,0) = 1$. But what about $\Gamma(ds_1, ds_2)$? For the linear case we have

$$(1 + L_1 ds_1)(1 + L_2 ds_2) = \Gamma(ds_1, ds_2)(1 + L_2 ds_2)(1 + L_1 ds_1) \,. \qquad \text{(A62a)}$$

With

$$\Gamma(ds_1, ds_2)_0 = 1 + (\partial^2 \Gamma / \partial s_1 \partial s_2)_0 \, ds_1 ds_2$$

$$(\partial^2 \Gamma / \partial s_1 \partial s_2)_o = L_1 L_2 - L_2 L_1 \qquad \text{(A63)}$$

follows.

The condition that the group commutator of the infinitesimal elements also be such an element

$$g(ds_1) g(ds_2) g^{-1}(ds_1) g^{-1}(ds_2) = g(ds_1, ds_2)$$

leads in the linear case to the equation

$$[L_1, L_2] = L' \qquad \text{(A64)}$$

where L' is a linear combination of the L_k. This is a necessary condition for the completeness of the set of n generators. The group of infinitesimal elements linear in L is the "local group". Eq. (A63) in conjunction with eq. (A64) is an integrability condition (Weyl 1931).

The Baker-Campbell-Hausdorff formula tells us that the product of two group elements given as exponential maps involves the generators and their commutators. So the condition (A64) is also sufficient: a set of quantities whose linear combinations and whose commutators also belong to the set forms a "Lie

Algebra".

The completeness of a set of generators which forms a Lie algebra is obviously of tremendous importance. When applied to physics it means there is a complete set of quantum numbers.

We know how to reach a finite or global group element with the usual infinitesimal ones,

$$\lim_{n\to\infty}(1+L\,ds)^n = \lim_{n\to\infty}(1+(s/n)L)^n = \exp(sL) \ .$$

But what if one were to use the somewhat unusual infinitesimal element $\Gamma = (1+[L_1,L_2]ds_1 ds_2)$? Use $[L_1,L_2] = L$, then

$$\lim_{n\to\infty}(1+(s_1 s_2/n^2)L)^n = (1+(s_1 s_2/n)L)$$

as one immediately finds with the binomial theorem. Consequently,

$$\lim_{n\to\infty}(1+(s_1 s_2/n^2)L)^{n^2} = \exp(s_1 s_2 L) \ . \qquad \text{(A65)}$$

This is in agreement with the expression for the second derivative of Γ, see eq. (A63).

In conclusion the basics of "Lie Groups", continuous groups connected to the identity: specifying a group element by the values of its n independent parameters, the group multiplication is written

$$g(\alpha_k) = g(\beta_k)\,g(\gamma_k)$$

and the inverse element

$$g(\delta_k) = g^{-1}(\alpha_k) \ .$$

For a Lie group

$$\alpha_k = \alpha_k(\beta_l;\gamma_l)$$
$$\textit{and}\ \delta_k = \delta_k(\alpha_l)$$

are analytic functions. The groups in this text are Lie groups.

REFERENCES

Adler, R., Bazin, M. and Schiffer, M. (1975), Introduction to General Relativity, McGraw-Hill Book Co., 2nd Ed., p. 275.

Adler, S. L. (1965), Phys. Rev. 139B: 1638.

Agrawal, V. et al., (1996) Phys. Rev. D 53: 1313.

Ahrens, T. (1987), Nuovo Cim. 98A: 25; (1989), Nuc. Phys. B6: 124; (1984), Interntl. J. Theoret. Phys. 23: 185; with Lang, T. P. (1971), Phys. Rev. C3: 979; with Gallaher, L. (1979), Phys. Rev. D20: 2714.

Aitchison, I. J. R. and Hey, A. J. G. (1989), Gauge Theories in Particle Physics, Adam Hilger, p. 416, p. 278.

Anselmann, P. et al. (1995), Phys. Lett. B342: 440.

Apollonio, M. et al. (1998), Phys. Lett. B420: 397.

Arnison, G. et al. (1983), Phys. Lett. B122: 103.

Athanassopoulos, C. et al. (1996), Phys. Rev. C54: 2685 (1998), Phys. Rev. Lett. 81: 1774.

Avignone, F. T. (1994), Prog. Particle and Nucl. Phys. 32: 223; with Brodzinski, R. L. (1988), Prog. Particle and Nuc. Phys. 21: 99.

Babu, K. S. (1988), Phys. Lett. B 203: 132.

Backe, H. et al. June (1996), Neutrino '96, Helsinki, World Scientific, p. 259.

Bagnaia, P. et al. (1983), Phys. Lett. 129B: 130.

Bahcall, J. N. (1970), Neutrinos from the Sun, Frontiers in Astronomy, Scientific Am., p. 108; with Basu, S. A., Pinsonneault, M. H. (1998), Phys. Lett. B433:1; with Krastev, P. I. and Smirnov, A. Yu (1998), Phys. Rev. D58: 1; with Primakoff, H. (1978), Phys. Rev. D18: 3463.

Banner, M. et al. (1983), Phys. Lett. B122; 476.

Barger, V., Weiler, T. J. and Whisnant, K. (1998), Phys. Lett. B 440: 1.

Barnes, V. et al. (1964), Phys. Rev. Lett. 12: 204.

Baudis, L. et al. (1999), Phys. Rev. Lett. 83: 41.

Behrends, R. E. et al. (1962), Rev. Mod. Phys. 34: 1.

Bemporad, C. (1999), Nuc. Phys. B (Proc. Suppl.) 77: 159.

Benvenuti, A. et al. (1974), Phys. Rev. Lett. 32: 800.

Bernatowicz, T. et al. (1992), Phys. Rev. Lett. 69: 2341.

Bilenki, S. M., Hosek and Petkov, S. T. (1980) Phys. Lett. B 94: 495.

Boehm, F. and Vogel, P. (1987), Physics of Massive Neutrinos, Cambridge, Univ. Press, p. 22, p. 95, p. 106, p. 117.

Bogoliubov, N. N. and Shirkov, D. V. (1959), Introduction to the Theory of Quantized Fields, transl. by Volkoff, G. M., Intersc. Pubs. Inc., p. 97, p. 46.

Brueckner, K. A. (1952), Phys. Rev. 86: 106.

Burkhardt, H. et al. (1988), Phys. Lett. B206:169.

Byers, N. and Peierls, R. E. (1958), Nuovo Cim. 10: 520.

Cabibbo, N. (1963), Phys. Rev. Lett. 10: 531.

Cavaignac, J. F. et al. (1984), Phys. Lett. B148: 387.

Cheng, T. P. and Li, L. F. (1980), Phys. Rev. D 22: 2860.

Chibisov, G. B. (1976), Sov. Phys. Usp. 19: 624.

Christenson, J. H. et al. (1964), Phys. Rev. Lett. 13: 138.

Cini, M. and Gamba, A. (1953), Nuovo Cim. 10: 1040.

Cleveland, B. T. et al. (1998), Astrophys. J. 496: 505.

Corson, E. M. (1953), Introduction to Tensors, Spinors, and Relativistic Wave-Equations, Hafner Pub. Co., p. 18, p. 179, p. 181.

Cowan, C., Reines, Harrison, Cruse and McGuire (1956), Science 124: 103; Nature (1956), 178: 446.

Danby, G. et al. (1962), Phys. Rev. Lett. 9: 36.

Davis, L. Jr. et al. (1975), Phys. Rev. Lett. 35: 1402.

Davis, R. Jr., Harmer, D. S., and Hoffman, K. C. (1968), Phys. Rev. Lett. 20: 1205.

Duffin, R. J. (1938), Phys. Rev. 54: 1114.

Elliott, S. R., Hahn, A. A., and Moe, M. K. (1987), Phys. Rev. Lett. 59: 2020.

Englert, F. (1964), Phys. Rev. Lett. 13: 321.

Feldman, G. (1977), Uncertainty Principle and Foundations of Quantum Mechanics: A Fifty Years' Survey, eds. Wm. C. Price and S. S. Chissick, John Wiley and Sons, Inc., p. 365.

Fermi, E. (1934), Z. Physik 88: 161; (1950), Nuclear Physics, University of Chicago, p. 201.

Feynman, R. P. (1962), The Theory of Fundamental Processes, W. A. Benjamin Inc., p. 73; with M. Gell-Mann (1958), Phys. Rev. 109: 193.

Fiorini, E. et al. (1967), Physics Letters 25B: 602.

Fonda, L. and Ghirardi, G. C. (1970), Symmetry Principles in Quantum Mechanics, Marcel Decker, Inc., New York, p. 156.

Frauenfelder, H. (1957), Phys. Rev. 106: 386.

Fronsdal, C. (1963), 1962 Brandeis Lectures, Vol. 1: 427.

Fukuda, Y. et al. (1996), Phys. Rev. Lett. 77: 1683; (1998), Phys. Rev. Lett. 81: 1158 and 1562; Phys. Lett. B 436: 33; (1999), Phys. Rev. Lett. 82: 2430.

Furry, W. H. (1939), Phys. Rev. 56: 1184.

Gaisser, T. K. et al. (1996), Phys. Rev. D 54: 5578.

Garvin, V. et al. (1997), Neutrino '96, Helsinki, World Scientific, Singapore, p.14.

Gell-Mann, M. (1964), Phys. Lett. 8: 214; (1953), Phys. Rev. 92: 833; (1962), Phys. Rev. 125: 1067; (1962), Proc. International Conf. High Energy Physics, CERN, p. 805; with Pais, A. (1955), Phys. Rev. 97: 1387.

Gelmini, G. B. and Roncadetti, M. (1981), Phys. Lett. 99 B: 411.

Gerstein, S. S. and Zeldowich, Y. B. (1955), JETP (USSR) 29: 698; (1957), Sov. Phys. JETP 2: 576.

Glashow, S. L. (1961), Nuc. Phys. 22: 579; with Iliopoulos, J. and Maiani, L. (1970), Phys. Rev. D 2: 1285; (1975), Quarks with Color and Flavor, Scientific American 233,4: 38.

Goeppert-Mayer, M. (1935), Phys. Rev. 48: 512.

Goldberger, M. L. and Treiman, S. B. (1958), Phys. Rev. 110: 1178 and 111: 354.

Goldhaber, M., Grodzins, L. and Sunyar, A. (1958), Phys. Rev. 109: 1015.

Goldstone, J. (1961), Nuovo Cim. 19: 154.

Greenberg, O. W. (1964), Phys. Rev. Lett. 13: 598.

Gribov, V. and Pontecorvo, B. (1969), Phys. Lett. 28 B: 493.

Grimus, W., Pfeiffer, R. and Schwetz, T. (1999), UWThPh-1999-36, Vienna.

Gurr, H. S., Reines, F. and Sobel, H. W. (1974), Phys. Rev. Lett. 33: 179.

Hampel, W. et al. (1995), Phys. Lett. B 388: 364.

Han, D., Kim, Y. S., and Son, D. (1982), Phys. Rev. D 26: 3717; (1986), J. Math. Phys. 27: 2228; et al. (1984), Am. J. Phys. 52: 1037.

Hasert, F. J. et al. (1973), Phys. Lett. 46B: 138.

Haxton, W. C. and Stephenson, G. Y. Jr. (1984), Nuc. Phys. 12: 409.

Heisenberg, W. (1932), Z. Physik 77: 1.

Hayden, R. J. and Inghram, M. G. (1953), NBS Circ. 522: 189.

Higgs, P. W. (1964), Phys. Rev. Lett. 13: 508.

Hill, J. E. (1995), Phys. Rev. Lett. 75: 2654.

Hollweg, J. K. (1974), Phys. Rev. Lett. 32: 961.

Jackson, J. D. (1958), The Physics of Elem. Particles, Princeton Univ. Press, p. 104, p. 113.

Johnson, G. (1999), Strange Beauty, Alfred A. Knopf, New York, p. 207.

Kane, G. (1987), Modern Elementary Particle Physics, Addison-Wesley Pub. Co., p. 249.

Kayser, B. (1987), Proc. XXIII International Conf. High Energy Physics, World Scientific, Singapore.

Kemmer, N. (1939), Proc. Roy. Soc. A173: 91; (1938), Proc. Cambridge Phil. Soc. 34: 354.

King, R. W., and Perkins, J. F. (1958), Phys. Rev. 112: 963; private communication regarding Proc. Mid-West Conf. on Theoret. Phys. (St. Louis, Mo., March 1958).

Klapdor-Kleingrothaus, H. V. (1999), hep-ex/9907040.

Kobayashi, M. and Maskawa, K. (1973), Prog. Theor. Phys. 49: 652.

Konopinski, E. J. and Mahmoud, H. M. (1953), Phys. Rev. 92: 1045.

Kwon, H. et al. (1981), Phys. Rev. D27: 1228.

Kuzmin, V. A. (1966), Sov. Phys. JETP 22: 1050.

Lakes, R. (1998), Phys. Rev. Lett. 80: 1826.

Landau, L. D. (1954), Nuc. Phys. 3: 127.

Lang, T. P. et al. (1978), Conf. Proc. Neutrinos-78, Purdue Univ., p. C68; ref. 1 in this article is a collection of refs. to early works on neutral currents.

Langacker, P. et al. (1983), Phys. Rev. D 27: 1228.

Lee, T. D. and Yang, C. N. (1956), Phys. Rev. 104: 254; (1957), Phys. Rev. 105: 1671; with Oehme, R. and Yang, C. N. (1957), Phys. Rev. 106: 340.

Lee, W. et al. (1976), Phys. Rev. Lett. 37: 186.

Lobashev, V. M. (1996), Neutrino '96, Helsinki, World Scientific, p. 265; et al. (1999), Nuc. Phys. B (Proc. Suppl.) 77: 327.

Luders, G. (1957), Ann. Phys. 2: 1.

Majorana, E. (1937), Nuovo Cim. 14: 171.

Mann, W. A., Kafka, T., and Leeson, W. (1992), Phys. Lett. B 291: 200.

Mikheyev, S. P. and Smirnov, A.Yu. (1986), Nuovo Cim. C9: 17.

Mnich, J. (1999), The LEP Collaborations, International Europhysics Conf., Tampere, Finland.

Moe, M. and Vogel, P. (1994), Ann. Rev. Nuc. and Particles 44: 247.

Morrison, P. (1953), in Experimental Nuc. Physics II, editor Segre, E., John Wiley and Sons, Inc. p. 1.

Muehlhause, C. O. and Oleksa, S. (1957), Phys. Rev. 105: 1332.

Ne′eman, Y. (1961), Nuc. Phys. 26: 222; (1962), see Johnson, G.

Nishijima, (1954), Prog. Theoret. Phys. 12: 107; 13: 285.

NIST Ref. on constants, etc. (1998), http://physics.nist.gov/cuu/constants/index.html

Okubo, S. (1959), Nuovo Cim. 13: 292.

Oppenheimer, J. R. and Schwinger, J. S. (1939), Phys. Rev. 56: 106.
Pais, A. and Piccioni, O. (1955), Phys. Rev. 100: 1487.
Pasierb, E. et al. (1979), Phys. Rev. Lett. 43: 96.
Pauli, W. (1940), Phys. Rev. 58: 716; (1957), Nuovo Cim. 6: 204.
Perkins, D. H. (1987), Introduction to High En. Physics, Addison-Wesley Pub. Co., p. 146, p. 160, p. 220, p. 384.
Pontecorvo, B. (1957), Zh. Eksp. Theor. Phys. 33: 549; (1967), ibid. 53: 1717; (1958), Sov. Phys. JETP 6: 429; (1948), Chalk River Report DD205.
Primakoff, H., with Sharp, D. H. (1969), Phys. Rev. Lett. 23: 501; with Rosen, S. P. (1969), Phys. Rev. 184: 1925.
Proca, A. (1936), J. Phys. Radium 7: 347.
Racah, G. (1937), Nuovo Cim. 14: 322; (1942), Phys. Rev. 62: 438.
Reines, F. (1983), Nuc. Phys. A 396: 469; with Cowan, C. (1959), Phys. Rev. 113: 273.
Review of Particle Physics 1996 (1996), Phys. Rev. D 54: 65; ibid. 1998 (1998), The European Physical J. C: p. 103, p. 319.
Sachs, R. G. and Treiman, S. B. (1956), Phys. Rev. 103: 1545.
Sakata, S. (1956), Progr. Theoret. Phys. 16: 686.
Sakurai, J. J. (1973), Advanced Quantum Mechanics, Addison-Wesley Pub. Co., p. 62; (1958), Nuovo Cim. 7: 649.
Salam, A. (1957), Nuovo Cim. 5: 299; (1968), Elementary Particle Theory, ed. Svartholm, N., Almqvist, p. 367.
Schechter, J. and Valle, J. W. F. (1980), Phys. Rev. D 22: 2227; (1982), ibid. D25: 2951.
Serpe, J. (1952), Physica 18: 295.
Schiff, L. I. (1949), Quantum Mechanics, McGraw-Hill Book Co., p. 144, p. 226..
Schwitters, R. F. (1977), Fundamental Particles with Charm, Scientific American 237,4: 56.
Stueckelberg, E. C. G. (1938), Helv. Phys. Acta 11: 225, 299.
Sudarshan, E. C. G. and Marshak, R. E. (1957), Proc. Padua-Venice Conf. on Mesons and Recently Disc. Particles.
Swartz, M. L. (1999), hep-ex/9912026 v2, XIX Intl. Symp. on Lepton a. Photon Ints., Stanford U.
Takaoka, N. and Ogata, K. (1966), Z. Naturf. 21a: 84.
Taylor, J. C. (1958), Phys. Rev. 110: 1216.
Tiomno, J. and Wheeler, J. A. (1949), Rev. Mod. Phys. 21: 153.

Valle, J. W. F. (1999), hep-ph/9911224.

Varadarajan, V. S. (1974), Lie Groups, Lie Algebras, and Their Representations, Prentice Hall, p. 118, p. 122.

Weinberg, S. (1965), The Quantum Theory of Massless Particles, Lectures on Particles and Field Theory, Brandeis Summer Inst. in Theoretical Physics 2, Prentice Hall, p. 417, p. 455; (1967), Phys. Rev. Lett. 19: 1264; (1964), Phys. Rev. 133B: 1318; (1996), The Quantum Theory of Fields II, Cambridge Univ. Press, p. 308, p. 182; (1995), volume I, p. 303, p. 339.

Weisberger, W. I. (1966), Phys. Rev. 143: 1302.

Wentzel, G. (1949), Quantum Theory of Fields, translated by Houtermans, C. and Jauch, J. M., Interscience Publishers, p. 63.

Weyl, H. (1946), The Classical Groups, Princeton Univ. Press, p. 265; (1929), Z. F. Physik 56: 330; (1931), The Theory of Groups and Quantum Mechanics translated by Robertson, H. P., Dover Publications Inc., p. 148, p. 177.

Wigner, E. P. (1939), Ann. Math. 40: 149; (1932), Gottinger Nachrichten 31: 546; (1931), Gruppentheorie u ihre Anwendung auf die Quantenmechanik der Atomspektren, Friedr. Viehweg und Sohn, Braunschweig, p. 99, p. 198; (1963), Theoret. Phys., Interntl. En. Agency, Vienna, p. 59.

Wolfenstein, L. (1978), Phys. Rev. D 17: 2369; with Lande, K. (1995), Symmetries and Fundamental Ints. in Nuclei, World Scientific, p. 307; (1981), Phys. Lett. B107: 77.

Woods, M. et al. (1988), Phys. Rev. Lett. 60: 1695.

Wu, C. S. E. and Ambler, Hayward, Hoppes, and Hudson (1957), Phys. Rev. 105: 1413.

Yang, C. N. and Mills, R. L. (1954), Phys. Rev. 96: 191. With Tiomno, J. (1950), Phys. Rev. 79: 495.

Young, H. D. (1968), Fundamentals of Optics and Modern Physics, McGraw-Hill, p. 51.

Yukawa, H. (1935), Proc. Phys.-Math. Soc. Japan 17: 48.

Zee, A. (1980), Phys. Lett. B 203: 132.

Zweig, G. (1964), CERN Report 8182/Th. 401.

INDEX